ASTRONOMY

Second Edition

More than 80 Self-Teaching Guides teach essential skills from management to math, computer programming to personal finance, writing to art appreciation.

Look for these and other popular STGs at your favorite bookstore!

Kuhn, BASIC PHYSICS
Houk & Post, CHEMISTRY: CONCEPTS AND PROBLEMS
Gordon, HOW TO SUCCEED IN ORGANIC CHEMISTRY
Kybett, ELECTRONICS
Sutton, ECOLOGY
Pearson, MATH SKILLS FOR THE SCIENCES
Gilbert & Gilbert, THINKING METRIC, 2nd Edition
Rothenberg, FINITE MATH
Kleppner & Ramsey, QUICK CALCULUS
Koosis, STATISTICS, 2nd Edition
Carman & Carman, QUICK ARITHMETIC
Locke, MATH SHORTCUTS
Adams, READING SKILLS
Gilbert, CLEAR WRITING
Markgraf, PUNCTUATION
Romine, VOCABULARY FOR ADULTS
Ryan, SPELLING FOR ADULTS
Carman & Adams, STUDY SKILLS: A STUDENT'S GUIDE FOR SURVIVAL
Lolley, YOUR LIBRARY: WHAT'S IN IT FOR YOU?
Grossman, QUICK TYPING
Grossman, QUICKHAND
Bell, Hess, & Matison, ART: AS YOU SEE IT
Hess, APPRECIATING LITERATURE: AS YOU READ IT
Seyer, Novick, & Harmon, WHAT MAKES MUSIC WORK
Ferner, SUCCESSFUL TIME MANAGEMENT
Jongeward & Seyer, CHOOSING SUCCESS: T.A. ON THE JOB
Zimmerman, MANAGING YOUR OWN MONEY
Albrecht, Finekl, & Brown, BASIC, 2nd Edition
Albrecht, Finkel, & Brown, BASIC FOR HOME COMPUTERS
Ashley, BACKGROUND MATH FOR A COMPUTER WORLD, 2nd Edition

ASTRONOMY

SECOND EDITION

Dinah L. Moché, Ph.D.

Professor
Department of Physics
Queensborough Community College
of the City University of New York
Bayside, NY 11364

Star Maps by
George Lovi
Sky and Telescope
Cambridge, Massachusetts

John Wiley & Sons, Inc.
New York • Chichester • Brisbane • Toronto • Singapore

Publisher: Judy V. Wilson
Editor: Dianne Littwin
Astronomy art by John King
Line art by Douglas Luna

520
M688a
c-1
7.95
10/82

Copyright © 1978, 1981 by John Wiley & Sons, Inc.

All rights reserved. Published simultaneously in Canada

Reproduction or translation of any part of this work beyond that permitted by Sections 107 or 108 of the 1976 United States Copyright Act without the permission of the copyright owner is unlawful. Requests for permission or further information should be addressed to the Permissions Department, John Wiley & Sons, Inc.

Library of Congress Cataloging in Publication Data

Moché, Dinah L., 1936-
Astronomy.

(Wiley self-teaching guides)
Bibliography: p.
Includes index.
1. Astronomy. I. Lovi, George. II. Title.
III. Series.
QB45.M696 1981 520 81-10470
ISBN 0-471-09713-6 AACR2

Printed in the United States of America

82, 81 10 9 8 7 6 5 4 3 2

CREDITS

Photographs

Photographs are courtesy of the following organizations and individuals:

C.S.I.R.O.—Fig. 2.15; Hale Observatories—Figs. I.4, 2.12(i), 3.3, 3.15, 4.8, 4.10, 5.1, 5.2, 5.3, 5.4, 5.6, 5.12, 5.16(c, d, e), 5.19, 5.20, 6.2, 7.10, 9.23, 9.24, 11.1, 11.3, 11.5; Harvard College Observatory—Fig. 7.8; Kitt Peak National Observatory—Figs. 2.6, 4.1, 5.5, 5.21, 7.7; Leiden Observatory—Fig. 5.7; Lick Observatory—Figs. 3.7, 9.22; Lowell Observatory—Fig. 9.20; NASA—Figs. I.1, I.2, I.3, 2.12(ii), 4.12, 7.3, 7.12, 7.13, 8.1, 8.2, 8.14, 8.15, 8.16, 9.1, 9.2, 9.6, 9.8, 9.14, 9.15, 9.16, 9.17, 9.18, 9.19, 9.21, 9.22, 10.1, 10.5, 10.6, 10.7, 11.2, 11.4, 11.9, 12.2, 12.3, 12.4; Princeton University Project Stratoscope—Fig. 7.9; J. William Schopf, Elso S. Barghoorn, Morton D. Maser, and Robert O. Gordon—Fig. 12.1; Dr. Martin Schwarzchild, Princeton University—Fig. 7.11; Tass/Sovfoto—Fig. 9.7; U.S. Air Force—Fig. 11.8; Yerkes Observatory, University of Chicago—Figs. 3.8, 3.9, 5.16(a, b, f), 5.18, 5.19, 7.5.

Illustrations and Tables

Illustrations and tables are adapted, redrawn, or used by permission of the following authors and publishers:

Fig. 5.11 (redrawn), Table 5.2, Table 10.2 (adapted)—from *Realm of the Universe*, by George O. Abell. Copyright © 1964, 1969, 1973, 1980, by Holt, Rinehart & Winston, Inc., copyright © 1976 by George O. Abell. Used by permission of Holt, Rinehart & Winston, Inc.

Tables 1.1 (columns 6-8), 3.1 (adapted), 3.4 (selected), 11.1 (selected), 11.2 (selected), 11.3—from *Astrophysical Quantities*, 3rd edition, © 1973, C. W. Allen (Athlone Press, London).

Figs. 1.1, 1.7—reprinted with permission of Edmund Scientific Co., publishers of *All About Telescopes*, by Sam Brown.

Fig. 1.5—reprinted with permission of Edmund Scientific Co., publishers of *All About Telescopes*, by Sam Brown, and *Rotating Star and Planet Locator*, by George Lovi.

Table 1.1 (columns 1-5), Figs. 1.11, 1.12, 2.1, 3.8 (assembly adapted), 3.12, 4.4, 4.5, 4.7, 5.14, 5.15, 5.17, 6.4, 6.5, 7.6, 7.15, 8.3, 8.7 (adapted), Figs. 9.10, 10.12, 10.13, 10.14—redrawn with permission from *Astronomy: Fundamentals and Frontiers*, 3rd edition, by Robert Jastrow and Malcolm H. Thompson (John Wiley & Sons, New York). © 1972, 1974, 1977 by Robert Jastrow.

Table 8.1 adapted and selected from NASA public information.

Fig. 10.2—redrawn by permission from *Introductory Astronomy*, by Nicholas A. Pananides, © 1973, Addison-Wesley, Reading, Massachusetts.

Figs. 6.1, 6.3 (adapted), 6.7 (adapted), 8.11—redrawn with permission from *Astronomy: A Multimedia Approach*, by William Protheroe et al. Copyright © 1976 by Charles E. Merrill Publishing Company.

Figs. 9.3, 9.4, 9.5, 9.9, 10.4—adapted from *Astronomy: The Evolving Universe* by Michael Zeilik. Copyright © 1976, 1979 by Michael Zeilik. By permission of Harper & Row, Publishers, Inc.

CAMBRIA COUNTY LIBRARY
JOHNSTOWN, PA. 15901

Acknowledgments

I am especially grateful to my numerous students and lecture audiences and to readers of *Astronomy* whose questions and comments formed the basis of the second edition.

Many people have encouraged, influenced, and supported my work, I thank:

Harry L. Shipman at the American Astronomical Society; Lloyd Motz and Chien Shiung Wu at Columbia University; Frank E. Bristow (JPL), Les Gaver, David W. Garrett, Curtis M. Graves, William D. Nixon (Headquarters); Peter W. Waller (Ames), and Terry White (JSC) of the National Aeronautics and Space Administration; Janet K. Wolfe at the National Air and Space Museum; Richard W. West at the National Science Foundation; Henry D. Berney, Thomas Como, Donald Cotten, Julius Feit, Sheldon E. Kaufman, Valdar Oinas, Robert Taylor, and President Kurt R. Schmeller at Queensborough Community College; and Arnold A. Strassenburg at The State University of New York at Stony Brook.

My daughters Elizabeth and Rebecca Rozen who provided valuable editorial assistance and also helped test the manuscript for comprehensibility.

Publisher Judy V. Wilson, Irene F. Brownstone, and Joyce Campbell (first edition), and Doreen Jasquith and Dianne Littwin (second edition) at John Wiley & Sons, Inc.

Carol R. Leven, my secretary.

My parents Mollie and Bertram A. Levine and husband Leonard who have been a wonderful inspiration.

The National Science Foundation Faculty Fellowship in Science awarded to me made possible advanced studies in astronomy.

To the Reader

Astronomy is a self-instructional book with a unique approach to teaching this fascinating subject. It is designed so that students with no formal astronomy background can easily learn basic principles and concepts plus exciting contemporary topics by themselves at a pace they enjoy. The book can be used alone as an introduction to astronomy and space exploration or it may be used as a supplement to any of the many excellent conventional textbooks or methods of instruction currently in use. The topics selected for presentation are those most often taught in a one semester or quarter college level course in introductory astronomy.

While keeping its predecessor's successful design, the second edition has important additions, revisions, and improvements. The entire book has been updated. This edition incorporates *current* discoveries, tools, research, and scientific insights, as well as suggestions from many who have profitably used the first edition.

The chapters on the solar system have been rewritten to include brand-new scientific results from the historic Viking to Mars, Voyager to Jupiter and Saturn, Pioneer and Venera to Venus, Landsat to Earth, and Solar Maximum Missions. A new Appendix describes the properties and discovery of the known satellites of the solar system. Recent photographs show space scenes never before viewed by humans. Astronomical symbols are keyed. The Index has been expanded so the text can serve as a contemporary glossary of astronomy and space exploration.

There are several reasons why you will find this text particularly useful. The writing is lively and easy to understand. There are many clear drawings and photographs to help make technical ideas understandable. Mathematics is not required. Special star maps are provided so that you can go right outside and see real examples in the sky of many celestial objects discussed in the text. Simple experiments using common materials are included so that you can become more familiar with basic principles and concepts by testing these ideas at home.

You are constantly and actively *involved* in learning astronomy. You can see what information is contained in the text in the list of objectives. You test your understanding of the material in the text frequently. And you learn to identify interesting sky objects.

The material in each chapter is presented in short, numbered sections called *frames.* Each frame contains new information and usually asks you to answer a question, or asks you to make sure you understand the material by suggesting an explanation, analyzing, or summarizing as you go along. You should try to answer each question by yourself while keeping the book's answer covered. Then compare what you have written with the answer printed below the dashed line. If your answer agrees with the book's, you understand the material and are ready to proceed to the next frame. If it does not, you should review some previous frames to make sure you understand the material before you proceed.

At the end of each chapter you will find a self-test designed to let you find out how well you understand the material in the chapter. You may test yourself right after completing a chapter, or you might take a break and then take the self-test as a review before beginning a new chapter. Compare your answers with the book's. If your answers do not agree with the printed ones, review the appropriate frames (listed next to each answer). If you want to read more extensively about any topic you can refer to the appropriate chapter in one of the popular introductory astronomy textbooks listed in the Cross-Reference Chart.

Special lists of Useful References and Selected Resources for Astronomy Materials are included if you want to become more involved with astronomy in your leisure time or career. Here you will find selected magazines, observing guides and star atlases, general reference books, career information, sources of excellent, beautiful astronomy materials, guides to planetariums and observatories, and teaching materials. You will find further useful information in the five Appendixes.

The author and publisher have tried to make this book accurate, up-to-date, enjoyable, and useful for you. It has been read by astronomers and students who have contributed helpful suggestions during the preparation of the final manuscript. If, after completing the book, you have suggestions to improve it for future readers, please write to:

Editor, Self-Teaching Guides
John Wiley & Sons, Inc.
605 Third Avenue
New York, NY 10016

Cross-Reference Chart

The chart on page xii shows the chapters in some popular astronomy textbooks that treat the material covered in each chapter of *Astronomy*. You may find this chart useful for additional reading on specific topics or in relating work in this book to a college course that uses one of these books. A bibliography of the books included in the chart is given below.

Abell, George O., *Realm of the Universe*, 2nd ed. (Holt, 1980).

Berman, Louis, and J.C. Evans, *Exploring the Cosmos*, 3rd ed. (Little, Brown, 1980).

Brandt, John C., and Stephen P. Maran, *New Horizons in Astronomy*, 2nd ed. (Freeman, 1979).

Fredrick, Laurence W., and Robert H. Baker, *An Introduction to Astronony*, 9th ed. (Van Nostrand, 1980).

Goldsmith, Donald, *The Evolving Universe*, 2nd ed. (Benjamin, 1981).

Hartmann, William K., *Astronomy, The Cosmic Journey,* 2nd ed. (Wadsworth, 1982).

Hodge, Paul W., *Concepts of Contemporary Astronomy* (McGraw-Hill, 1979).

Jastrow, Robert, and Malcolm H. Thompson, *Astronomy: Fundamentals and Frontiers*, 3rd ed. (Wiley, 1977).

Jefferys, William H., and R. Robert Robbins, *Discovering Astronomy* (Wiley, 1981).

Kaufmann, William J. III, *Astronomy: The Structure of the Universe* (Macmillan, 1977).

Motz, Lloyd, and Anneta Duveen, *Essentials of Astronomy*, 2nd ed. (Columbia University Press, 1977).

Pasachoff, Jay M., *Contemporary Astronomy*, 2nd ed. (Saunders, 1981).

Seeds, Michael A., *Horizons*, (Wadsworth, 1981).

Shipman, Harry, *The Restless Universe* (Houghton Mifflin, 1978).

Wyatt, Stanley P., *Principles of Astronomy, A Short Version*, 2nd ed. (Allyn & Bacon, 1981).

Zeilik, Michael, *Astronomy: The Evolving Universe*, 2nd ed. (Harper & Row, 1979).

	Abell	Berman, Evans	Brandt, Maran	Fredrick, Baker	Goldsmith	Hartmann	Hodge
Introduction: Cosmic View	1	1	1	1	1	I	1
1. Understanding the Starry Sky	2,4		2	1,2,3,11	2	1	
2. Light and Telescopes	5,6,7	4,5	5,6,14	4	3	13	
3. The Stars	12,13,14	12,13	10	12,13	8	14,18	8
4. Stellar Evolution	17,18	14,15	11,16	14,15	9,10	15,16,17	9,10,11, 12,13,14, 15,16,19
5. Galaxies	15,20	16,17	12,13	16,17,18	5,6,7	19,20,21, 22	18,19,20
6. The Universe	21	18	13	19	4	23,24	21,22,23 24
7. The Sun	16	11	9	10	8	12	11
8. Understanding the Solar System	3,8	2,3,7	7	6	11	2,4,6,11	2
9. The Planets	10	6,8,9	3,8	7,8	12,13	3,7,8,9	3
10. The Moon	9	6	15	5	12	5	4
11. Comets, Meteors, and Meteorites	11	8,9	7	9	11	10	5,6
12. Life on Other Worlds?	19	10,19	4		14	25	7

Jastrow, Thompson	Jefferys, Robbins	Kaufmann	Motz, Duveen	Pasachoff	Seeds	Shipman	Wyatt	Zeilik
1	1	1		1	I	1	1	1
I	1,2,3,4	1,3	1,2,5	4	1,2	2	3	1
2,3	10,11,12, 13	5,6	Appen- dices A & B,16	2	3,4	1	4	5,6
4,5	14,15,16 17	7	15,17,18, 19,20,21, 22,25,26	3,5,6	5	9	11,12,13	14
6,7,8	18,19	8,9,10,11	23,24	8,9,10,11	6,7	10,11,12	14,15,16	19,20
9,10,11	20,21,22, 23	12,13	27,28,29, 30	22,23,24, 25,26	8,9,10	13,14	17,18	15,16,17 18
12	24	14	31,32	27	11	15	19	7,21
13,14	17	7	14,20	7		8	10	8
15	5,6	2	8,9,12	12	12,13	3	6,9	2,3,4,13
16,18,19	9	4	3,4,12	14,15,16, 17,18,19	14,15	4,5,6	2,7,8	9,11,12
17	8	4	6,7	13	14	4	5	10
15	7	4	13	20	12	7	8	13
20				21	16	16	20	22

List of Tables

Contents

INTRODUCTION

Cosmic View

> Strange is our situation here upon Earth.
> Each of us comes for a short visit, not
> knowing why,
> Yet sometimes seeming to divine a purpose.
>
> Albert Einstein

On a clear night in a place where the sky is really dark, you can see about two thousand stars with your naked eye. You can look millions of miles out into space and peer far back into the distant past.

As you gaze at the stars you may wonder: What is the pattern or meaning to the starry heavens? What is my place in the vast cosmos? You are not alone in asking these questions. The beauty and mystery of space have always fascinated people.

Astronomy is the oldest science—and the newest. Exciting discoveries are being made today with the most sophisticated tools and techniques ever available. Yet dedicated amateurs can still make important contributions to astronomy.

This book will teach you the basic concepts of astronomy. You will more fully enjoy observing the stars as your knowledge and understanding of astronomy grow. You will be able to read more deeply into the literature on topics that intrigue you, from ancient astronomy to the newest astrophysical theories.

As you teach yourself astronomy, refer to:

★ The *star maps*, at the back of this book. These special, easy-to-read star maps will help you locate and identify particularly interesting objects in the sky.

Simple *experiments* you can do at home which demonstrate a basic idea.

Now, begin reading and stretch your mind.

Our home is the planet Earth, a tiny blue and white ball about 8,000 miles in diameter suspended in the vastness of space. Earth belongs to the solar system.

Figure I.1. Earth.

The solar system consists of one star—our sun—plus nine planets with their moons, thousands of asteroids, comets, and dust particles, all of which revolve around the sun. The solar system is about 8 billion miles across.

The sun and the solar system are located in one of the great spiral arms of the Milky Way Galaxy. The Milky Way Galaxy includes over 100 billion stars plus interstellar gas and dust, all revolving around the center. The Milky Way Galaxy is about 100,000 light-years (600 million billion miles) across.

Figure I.2. The solar system.

Figure I.3. The solar system in the Milky Way Galaxy.

Our Milky Way Galaxy is only one of millions of other galaxies that extend to the edge of the observable universe, some 12 billion light-years (72 billion trillion miles) away.

Figure I.4. Galaxies.

CHAPTER ONE

Understanding the Starry Sky

> And that inverted bowl we call the Sky
> Where under crawling coop't we live and die
> Lift not your hands to It for help—for It
> As impotently rolls as you and I.
>
> *Rubáiyát of Omar Khayyám*

1. On a clear dark night the sky looks like a gigantic inverted bowl studded with stars. We can easily see why the ancients believed that the starry sky was a huge sphere turning around Earth.

Today we know that stars are remote blazing suns racing through space at different distances from Earth. The earth *rotates*, or turns, around its *axis* (the imaginary line running through the center between the North and South Poles), daily.

But the picture of the sky as a huge, hollow, rotating globe with Earth at the center and sky objects on the inside surface is useful. Astronomers call

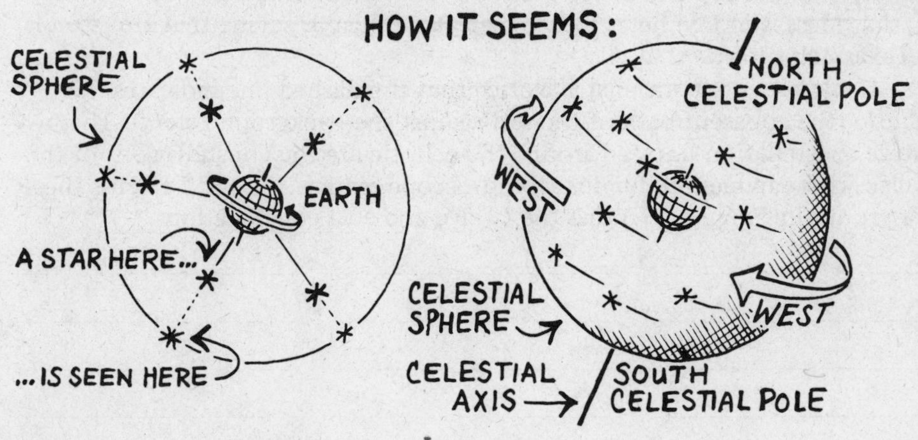

Figure 1.1. The celestial sphere.

this fictitious picture of the sky the *celestial sphere*. "Celestial" comes from the Latin word for heaven.

Astronomers use the celestial sphere to locate stars and galaxies and to plot the courses of the sun, moon, and planets throughout the year. When you look at the stars, imagine yourself inside the celestial sphere, looking out.

Why do the stars on the celestial sphere appear to move by you during the night when you observe them from Earth? ______________________

____

- - - - - - - - - - - - - - - - - -

Because the earth is rotating on its axis inside the celestial sphere.

2. ★ It is fun to go outside and see a young blue-white star or a dying red giant star in the sky right after you read about them. You may think you will never be able to tell one star from another when you begin stargazing, but you will. The star maps at the back of this book have been drawn especially for beginning stargazers observing from around 40° N latitude. (They should be useful to new stargazers throughout the continental United States.)

Stars appear to belong to groups that form recognizable patterns in the sky. These star patterns are called *constellations.* Learning to identify the most prominent constellations will help you pick out individual stars.

The eighty-eight officially recognized constellations are listed in Appendix 1. The more prominent, well-known ones that shine in these latitudes are shown on your star maps. Their names are printed in Latin in capital letters.

Thousands of years ago people named the constellations after animals (such as Leo the Lion) or mythological characters (such as Orion the Hunter). The ancient Greeks recognized forty-eight constellations more than 2,000 years ago.

Modern astronomers use the ancient names of the constellations to refer to eighty-eight sections of the sky rather than to the mythical figures of long ago. They locate sky objects by referring to constellations. For instance, saying that Mars is in Leo helps locate that planet just as saying that Houston is in Texas helps locate that city.

Look over your star maps. Notice that the dashed line indicates the *ecliptic* (the apparent path of the sun against the background stars). The twelve constellations located around the ecliptic are the constellations of the zodiac whose names are familiar to horoscope readers. (We will discuss these star groups in frame 1.17.) List the twelve zodiacal constellations.

______________ ______________ ______________

______________ ______________ ______________

______________ ______________ ______________

______________ ______________ ______________

- - - - - - - - - - - - - - - - - -

Pisces, Aries, Taurus, Gemini, Cancer, Leo, Virgo, Libra, Scorpius, Sagittarius, Capricornus, Aquarius

3. ★ Study your star maps carefully. You will notice that several of the constellations appear on all four maps. These *circumpolar constellations* are visible above the northern horizon all year long at around 40°N latitude.

List the three circumpolar constellations closest to Polaris (the North Star), and sketch their outlines. ______________________________

____

- - - - - - - - - - - - - - - - - -

Three circumpolar constellations that you should be able to pick out on the maps are Cassiopeia, Cepheus, and Ursa Minor. After you know their outlines, try to find them in the sky above the northern horizon. Note: At latitude 40°N or higher, Ursa Major and Draco are also circumpolar.

4. ★ Use the star maps out-of-doors to identify the constellations and stars you see in the night sky.

Choose the map that pictures the sky at the month and time you are stargazing. Turn the map so that the name of the compass direction you are facing appears across the bottom of the map. Your star map then pictures the sky as you are viewing it from your horizon to your *zenith* (the point directly overhead).

For example, if you are facing north about 10:00 P.M. in early April, turn the map so that the word NORTH is at the bottom. From the horizon up, you may observe Cassiopeia, Cepheus, the Little Dipper in Ursa Minor, and the Big Dipper in Ursa Major.

Name a prominent constellation that shines in the south at about 8:00 P.M. in early February. ______________________________

- - - - - - - - - - - - - - - - - -

Orion

5. ★ The constellations above the southern horizon parade by overnight and change with the seasons. Turn the map so that the word SOUTH is at the bottom. Use your star map to identify the most prominent constellations that shine each season (such as Leo in the spring and Orion in the winter).

Identify and sketch three constellations that you can see this season.

____

- - - - - - - - - - - - - - - - - -

Your answer will depend on the season. For example, if you are reading this book in the spring, you might choose Leo, Virgo, and Boötes.

6. ★ Long ago, more than fifty of the brighter stars were given individual names in Arabic, Greek, or Latin. The names of bright or famous stars are printed on your star maps in lowercase letters. Although many other stars have names also, astronomers usually refer to stars by Greek letters or numbers in star catalogues, not shown on your maps.

In a built-up metropolitan area you can see only the brightest stars. When you are far from city lights and buildings and the sky is very dark and clear, you can see about two thousand stars with your naked eye.

Name the three bright stars that mark the points of the famous summer triangle. Refer to your summer star map.

- - - - - - - - - - - - - - - - - -

Vega, Deneb, Altair. Look for the summer triangle overhead during the summer.

7. Some stars in the sky look brighter than others. The *magnitude* of a sky object is a measure of how bright it looks when viewed from Earth. Sky objects may look bright because they send out a lot of light or because they are very close to Earth.

In the second century B.C., the Greek astronomer Hipparchus divided stars into six classes, or magnitudes, by their relative brightness. He numbered the magnitudes from 1 (the brightest) through 6 (the least bright).

Modern astronomers use a similar classifying system. But instead of judging brightness with the naked eye, they use a photometer to measure it. Magnitudes for the brightest stars are negative—the brightest star, Sirius, measures −1.5 and magnitudes range to about +24 for the faintest objects seen in large telescopes. A difference of 1 magnitude means a brightness ratio of about $2\frac{1}{2}$. Magnitudes are indicated on your star maps.

For example, a star of magnitude 0, like Vega, looks about 2½ times as bright as a star of magnitude 1, like Deneb, and about 6.3 times as bright as Polaris, of magnitude 2. (Magnitudes are discussed further in frame 3.14.)

What do astronomers mean by "magnitude"? ____________________

- - - - - - - - - - - - - - - - - -

how bright a sky object looks

8. The more you understand about stars and their motions, the more you will enjoy stargazing. A *celestial globe* helps you locate sky objects as a *terrestrial* (earth) *globe* helps you locate places on Earth.

Remember how earth maps work. We picture the earth as a sphere and draw imaginary guidelines on it. All distances and locations are measured from two main reference lines, each marked 0°. One, the *equator*, is the great circle halfway between the North and South Poles that divides the globe into two equal halves. The other, the *prime meridian*, runs from pole to pole, through Greenwich, England.

Imaginary lines parallel to the equator are called *latitude* lines. Those from pole to pole are called *longitude* lines, or *meridians.* Distance on the terrestrial sphere is measured by dividing it into 360 sections called degrees (°).

You can locate any city on Earth if you know its coordinates of latitude and longitude.

Referring to figure 1.2, identify the equator, prime meridian, 30°N latitude line, and 30°E longitude line:

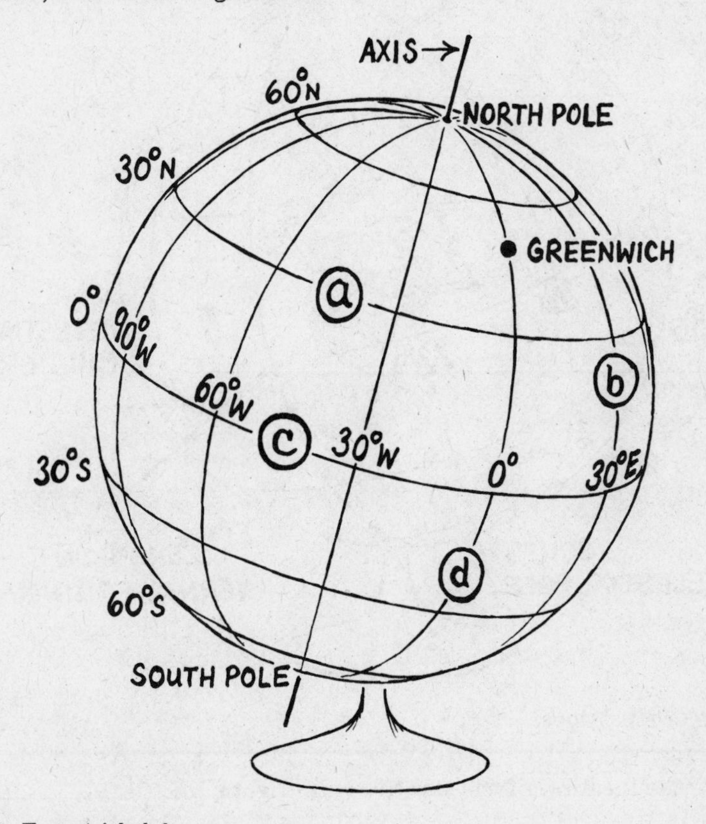

Figure 1.2. Terrestrial globe.

(a) ______________________; (b) ______________________;

(c) ______________________; (d) ______________________

— — — — — — — — — — — — — — — — — —

(a) 30°N; (b) 30°E; (c) equator; (d) prime meridian

9. Astronomers draw imaginary horizontal and vertical lines on the celestial sphere similar to the latitude and longitude lines on Earth. Angular distance above or below the celestial equator is called *declination* (dec). Distance measured eastward along the celestial equator from the zero point, the vernal

equinox, is called *right ascension* (RA). Right ascension is commonly measured in hours, with $1^h = 15°$.

Just as any city on Earth can be located by its coordinates of longitude and latitude, any sky object can be located on the celestial sphere by its coordinates of right ascension and declination.

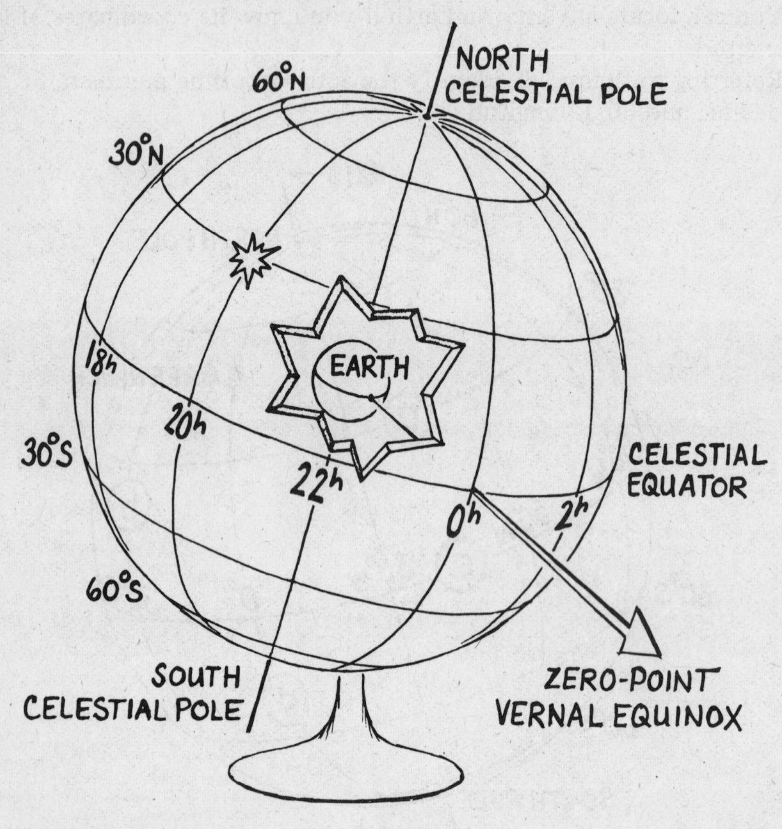

Figure 1.3. Celestial globe.

Give the location of the star shown in figure 1.3. ____________________

____

- - - - - - - - - - - - - - - - - - - -

20^h RA, 30°N dec

10. Every star has a location on the celestial sphere, where it appears to be when sighted from Earth. The right ascension and declination of stars change little over a period of many years and can be read from a celestial globe or star atlas. (See Table 1.1, for example. You'll be referring back to this table when the information it gives is explained in later chapters.)

Table 1.1. The twenty brightest stars in order of decreasing brightness

Name	Right Ascension Hours	Right Ascension Minutes	Declination Degrees	Declination Minutes	Distance (LY)	Spectral Class	Apparent Magnitude	Absolute Magnitude
Sirius	6	42.9	−16	39	8.8	A1	−1.5	+1.4
Canopus	6	22.8	−52	40	98	F0	−0.7	−4.7
Arcturus	14	13.4	+19	27	36	K2	−0.1	−0.2
Alpha Centauri	14	36.2	−60	38	4.3	G2	0.0	+4.4
Vega	18	35.2	+38	44	26	A0	0.0	+0.5
Capella	5	13.0	+45	57	46	G8	+0.1	−0.6
Rigel	5	12.1	− 8	15	900	B8	+0.1	−7.0
Procyon	7	36.7	+ 5	21	11	F5	+0.4	+2.7
Achernar	1	35.9	−57	29	150	B5	+0.5	−2.2
Hadar	14	0.3	−60	8	392	B1	+0.6	−5.0
Altair	19	48.3	+ 8	44	16	A7	+0.8	+2.3
Betelgeuse	5	52.5	+ 7	24	700	M2	+0.8	−6.0
Aldebaran	4	33.0	+16	25	68	K5	+0.9	−0.7
Alpha Crucis	12	23.8	−62	49	350	B2	+1.0	−3.5
Spica	13	22.6	−10	54	230	B1	+1.0	−3.4
Antares	16	26.3	−26	19	400	M1	+1.0	−4.7
Pollux	7	42.3	+28	9	35	K0	+1.2	+1.0
Fomalhaut	22	54.9	+29	53	23	A3	+1.2	+2.0
Deneb	20	39.7	+45	6	1,400	A2	+1.25	−7.3
Beta Crucis	12	44.8	−59	25	500	B0	+1.3	−4.7

The sun, moon, and planets have locations on the celestial sphere that change regularly. Their monthly positions are listed in current astronomical publications (see Useful References).

Can you explain why in any given era the stars may be found at the same coordinates on the celestial sphere, while the sun, moon, and planets change their locations regularly? ______________________________

____

____

— — — — — — — — — — — — — — — — — —

The stars are much too far away from Earth for the naked eye to see them move even though they are traveling many miles per second in various directions. The sun, moon, and planets are much closer to Earth. We see them move relative to the distant stars.

11. Lines of declination and right ascension are fixed in relation to the celestial sphere and move with it as it rotates around an observer. Other local reference lines relate to the position of each individual observer and stay fixed with the observer while sky objects pass by.

The *celestial horizon* is the great circle on the celestial sphere 90° from the observer's *zenith.* The zenith is the point on the celestial sphere directly over the observer's head. Although the celestial sphere is filled with stars, an observer can see only those that are above his horizon. The *celestial meridian* is the great circle passing through the zenith and the north and south points on the observer's horizon. (Only half of the celestial meridian is above the horizon.)

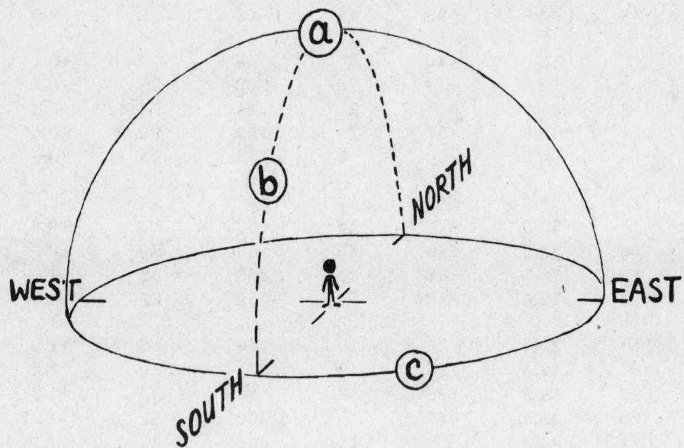

Figure 1.4. The stargazer's view.

Referring to figure 1.4, identify the stargazer's zenith, celestial horizon, and celestial meridian.

(a) ________________; (b) ________________; (c) ________________.

— — — — — — — — — — — — — — — — — — — —

(a) zenith; (b) meridian; (c) horizon

12. Go outside and trace out your zenith, celestial horizon, and celestial meridian by imagining yourself, like that stargazer, at the center of the huge celestial sphere.

If possible, try this on a clear, dark, starry night. Face south. Observe the stars near your celestial meridian several times during the night. Describe what you observe. ____

____

— — — — — — — — — — — — — — — — — — — —

The stars move from east to west and *transit* (cross) your celestial meridian. This is because of the earth's rotation from west to east.

13. The stars that appear above your horizon and their paths across the sky depend on your latitude on Earth. The sky looks different from different latitudes.

If you looked at the sky from the North Pole and then from the South Pole on the same night, you would see completely different stars. The earth cuts your view of the celestial sphere in half.

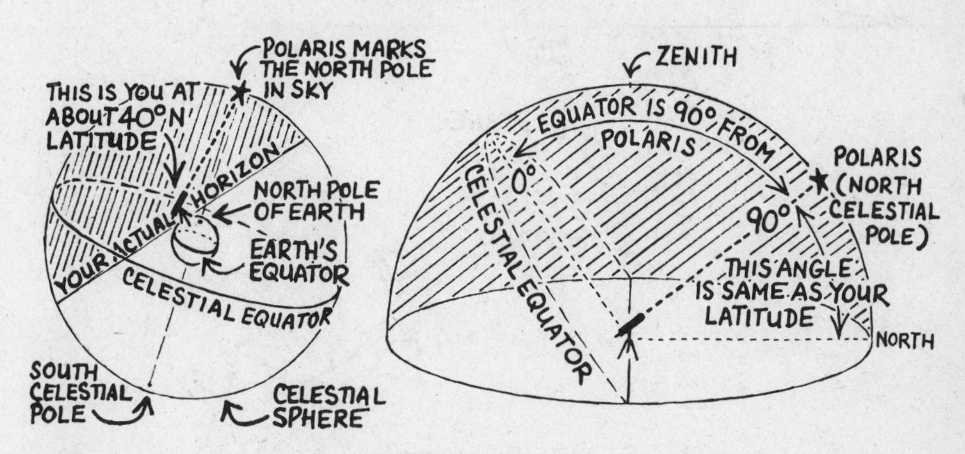

Figure 1.5. Local orientation of celestial sphere.

You can determine how the celestial sphere is oriented with respect to your horizon and zenith. The north celestial pole is located at an altitude above your northern horizon equal to your latitude above the equator. The position of the north celestial pole is marked in the sky by the North Star, Polaris, which is about 1° away from it.

Where would you look for the North Star if you were at each of the following locations:

(a) the North Pole? ____________; (b) the equator? ______________;

(c) 40°N latitude? ____________; (d) your home? ______________

____

- - - - - - - - - - - - - - - - - -

(a) at your zenith; (b) on your horizon; (c) 40° above your northern horizon; (d) at an altitude above your northern horizon equal to your home latitude

14. The North Star, Polaris, has long been important for navigation because of its location near the north celestial pole. It is not a very bright star. You can find Polaris by following the "pointer" stars, Dubhe and Merak, in the bowl of the Big Dipper.

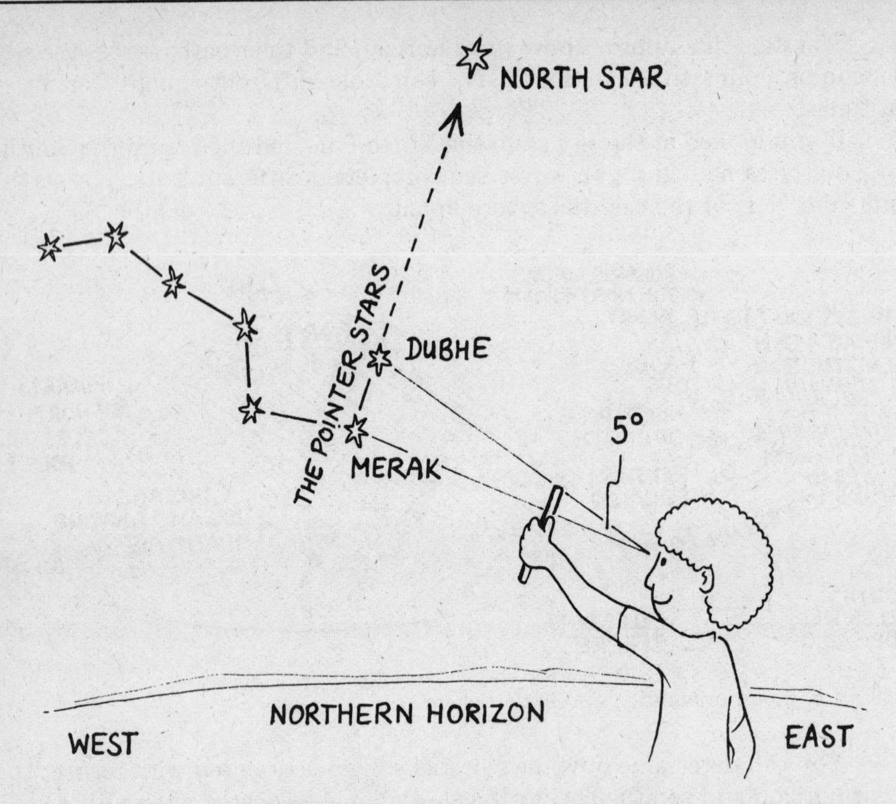

Figure 1.6. Finding the North Star and judging distance in the sky.

When you are stargazing, it is convenient to remember that the distance between these two pointer stars is about 5° on the celestial sphere. That fact will help you judge other sky distances with your naked eye.

All stars appear to move in *diurnal circles*, or daily paths, around the celestial poles when you observe them from the spinning earth. But the celestial poles are at different altitudes above your horizon at different latitudes. So the part of a star's diurnal circle that is above the horizon is different at different latitudes on Earth.

The diagram shows the stars' diurnal circles at 40°N latitude, about the latitude of New York City. The north celestial pole is 40° above the northern horizon, and the celestial sphere rotates around it. If you stargaze at 40°N latitude, you will see: (1) Stars within 40° (your latitude) of the north celestial pole (those stars between +50° and +90° declination) are always above your horizon. These stars that never set—such as the stars in the Big Dipper—are north *circumpolar stars.* (2) Stars that are within 40° (your latitude) of the south celestial pole—such as the stars in the Southern Cross—never appear above your horizon. (3) The other stars, in a band around the celestial equator, rise and set. Those stars that are located at 40°N declination (equal to your latitude) pass directly across your zenith when they cross your celestial meridian.

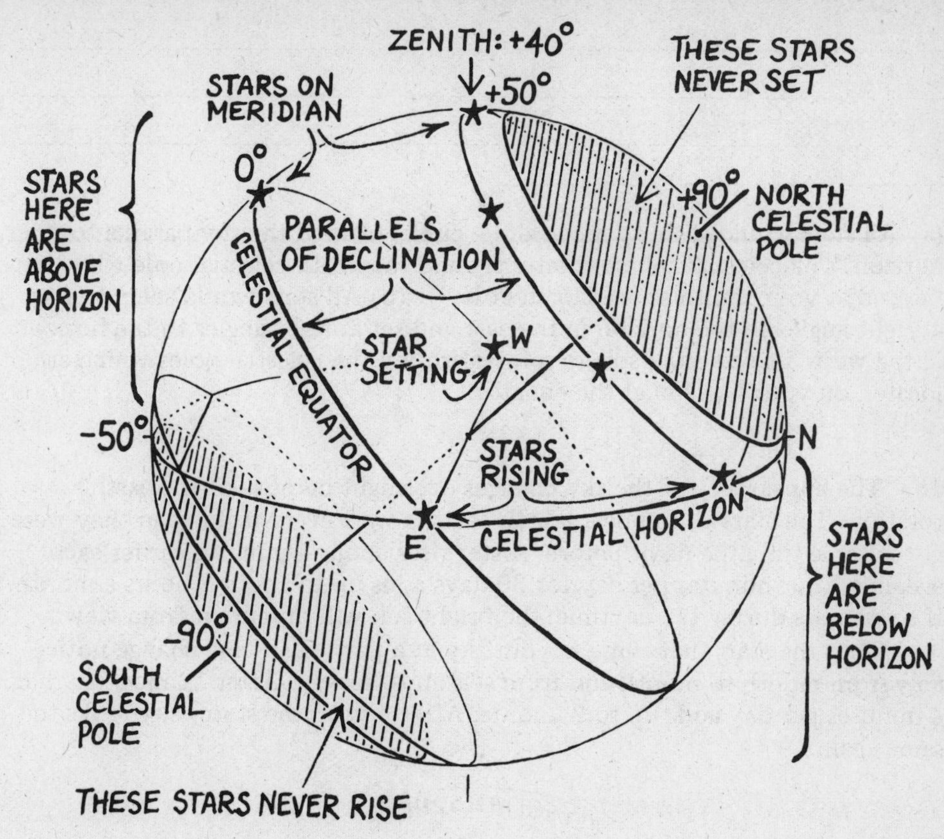

Figure 1.7. The sky from 40°N latitude.

Assume you are stargazing at 50°N latitude, about the latitude of Seattle, Washington. Refer to Table 1.1 for the declinations of the bright stars Capella, Vega, and Canopus. Which of these stars will be above your horizon:

(a) always ?__________; (b) sometimes?_______; (c) never? __________

– – – – – – – – – – – – – – – – – – – –

(a) Capella (+45°57′ dec). Stars within 50° of the north celestial pole (between +40° and +90°) are always above the horizon. (b) Vega (+38°44′ dec). This star rises and sets. (c) Canopus (−52°40′ dec) is within 50° of the south celestial pole (between -40° and -90°).

15. Describe how the diurnal circles of the stars would look if you were stargazing at (a) the North Pole; (b) the equator. Explain your answer. Note: Remember that the celestial sphere rotates around the celestial poles.

____

____

(a) All stars would seem to move along circles around the sky parallel to your horizon. The celestial sphere rotates around the north celestial pole which is located at your zenith at the North Pole. (b) All stars would seem to rise at right angles to the horizon in the east and set at right angles to the horizon in the west. The celestial sphere rotates around the celestial poles which are located on your horizon at the equator.

16. The appearance of the sky changes overnight because of the earth's rotation. The stars also appear a little farther west every night than they were at the same time the night before. A star rises about 4 minutes earlier each evening. Four minutes per day for 30 days adds up to about 2 hours a month. If a star rises during the daytime, the bright sun will obscure it from view.

Thus the stars that shine in your sky at a particular time change noticeably from month to month and from season to season. After 12 months, that 4 minutes per day adds up to 24 hours. After a year, the starry sky looks the same again.

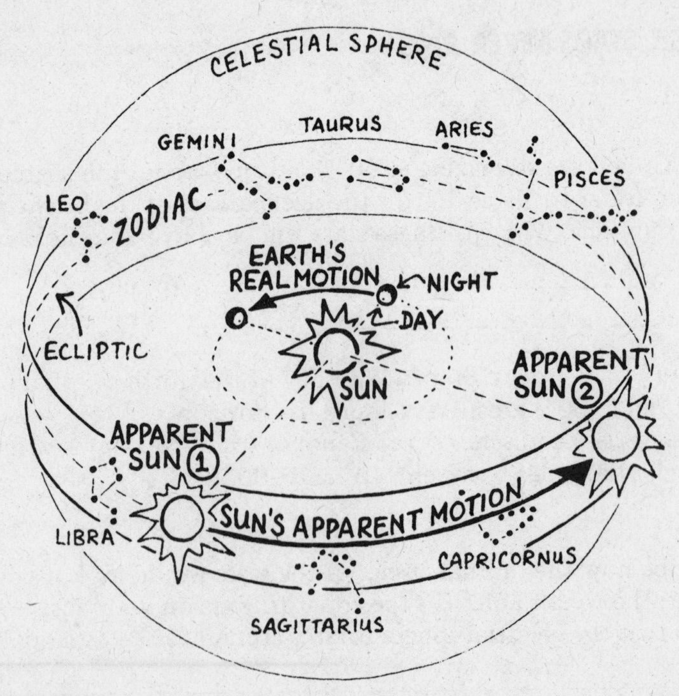

Figure 1.8. The ecliptic.

The change in the appearance of the sky with the change in seasons is due to the motion of the earth around the sun. The earth *revolves*, or travels around, the sun every year.

Picture yourself riding on Earth around the sun, inside the celestial sphere, looking straight out. As Earth moves along in its orbit, your line of sight points toward different stars in the night sky. During a whole year you view a full circle of stars.

(a) If a star is on your zenith at 9:00 P.M. on September 1, about what time will it be on your zenith on March 1? ___________

(b) Will you be able to see it? _________ Explain your answer.

____

____

– – – – – – – – – – – – – – – – – – – –

(a) about 9:00 A.M., since stars rise about 2 hours earlier every month
(b) No. At that hour of the day the bright sun obscures the distant stars from view.

17. If the stars were visible during the day you would see the sun apparently move eastward among them during the year. The *ecliptic*, the apparent path of the sun against the background stars, is drawn on sky globes and maps for reference (see, for example, your star maps).

The ecliptic passes through twelve constellations located on a belt about 16° wide around the sky. This belt of constellations was called the *zodiac* by ancient astrologers.

The zodiac attracted special attention because the moon and planets, when they appear in the sky, also follow paths near the ecliptic through these twelve constellations.

What is the zodiac? ______________________________________

____

– – – – – – – – – – – – – – – – – – – –

a belt about 16° wide around the sky, centered on the ecliptic, containing twelve constellations

18. The apparent easterly motion of the sun among the stars is caused by the real revolution of the earth around the sun. The sun seems to move in a full circle around the celestial sphere every year.

About how far does the sun move on the ecliptic every day? Hint: Use the fact that the sun moves 360° around the ecliptic in a year (about 365 days).

____

– – – – – – – – – – – – – – – – – – – –

about 1°. Solution: $\frac{360^\circ}{365 \text{ days}} \cong 1^\circ$ per day

19. The sun's path across the sky is highest in summer and lowest in winter. The altitude of the sun above the horizon at noon varies during the year because Earth's axis is tilted to the plane of its orbit around the sun. Earth's equator remains tilted at about $23\frac{1}{2}^\circ$ to the plane of its orbit around the sun all year long. So as Earth travels around the sun the slant of the Earth–sun line changes. Sunlight pours down to Earth from different angles, causing the change of seasons during the year, as well as seasonal variations in the length of days and nights.

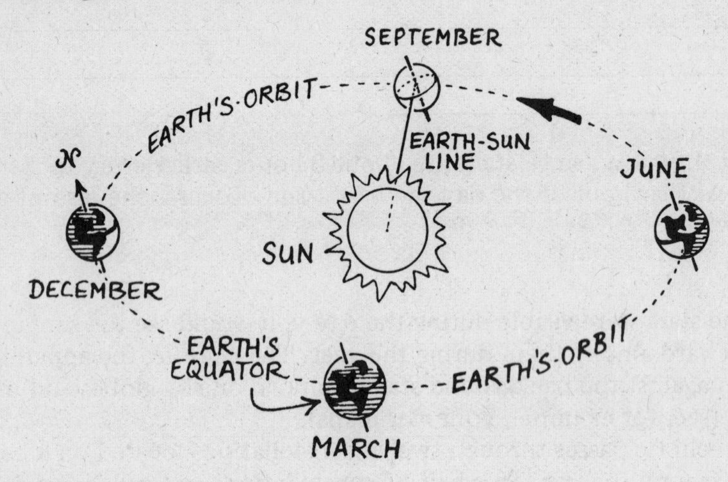

Figure 1.9. Earth's tilted axis and the seasons.

(a) In December, is the northern hemisphere tipped toward the sun or away? ______________ (b) In June, is the northern hemisphere tipped toward the sun or away? ______________

(a) away; (b) toward

20. You can determine what the sun's apparent position in the sky will be on any given day by checking the ecliptic on a celestial globe or a flat sky map like the one in figure 1.10.

The *vernal equinox* is the sun's position on about March 21 as it crosses the celestial equator going north. It is the point on the celestial sphere chosen to be the 0^h of right ascension (see frame 1.9). The *autumnal equinox* is about September 23, when the sun crosses the equator going south. At the equinoxes, day and night are equally long.

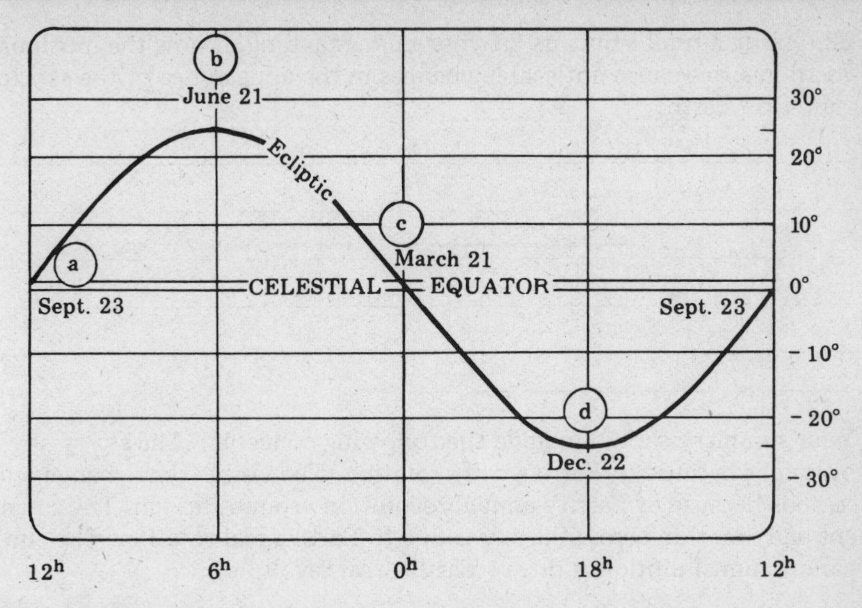

Figure 1.10. Flat sky map

The *summer solstice*, about June 21, and the *winter solstice*, about December 22, are the most northern and most southern positions of the sun during the year. At these times we have the longest and shortest days, respectively, in the northern hemisphere.

Referring to figure 1.10, identify the vernal equinox, autumnal equinox, summer solstice, and winter solstice.

(a) ______________________; (b) ______________________;

(c) ______________________; (d) ______________________

(a) autumnal equinox; (b) summer solstice; (c) vernal equinox;
(d) winter solstice

21. The sun is never directly overhead in the sky for stargazers in the United States. Where would you have to stand on Earth to have the sun pass directly across your zenith at the time of the

(a) vernal equinox? ______________ (b) summer solstice? ______________

(c) autumnal equinox? ______________ (d) winter solstice? ______________

(a) equator; (b) $23\frac{1}{2}$°N latitude (Tropic of Cancer); (c) equator;
(d) $23\frac{1}{2}$°S latitude (Tropic of Capricorn)

22. Write a brief summary of your understanding of how the motions of Earth in space cause noticeable changes in the appearance of the sky for a viewer on Earth.

Your summary should include the following concepts: The starry sky changes overnight because of Earth's daily rotation. The visible stars change with the seasons because of Earth's annual revolution around the sun. The sun's apparent daily motion across the sky is due to Earth's real rotation. The sun's apparent annual motion is due to Earth's real revolution.

23. Earth's rotation provides a basis for keeping time using astronomical observations. The *solar day* of everyday affairs measures the time interval of Earth's rotation using the sun for reference. The *sidereal day* measures the

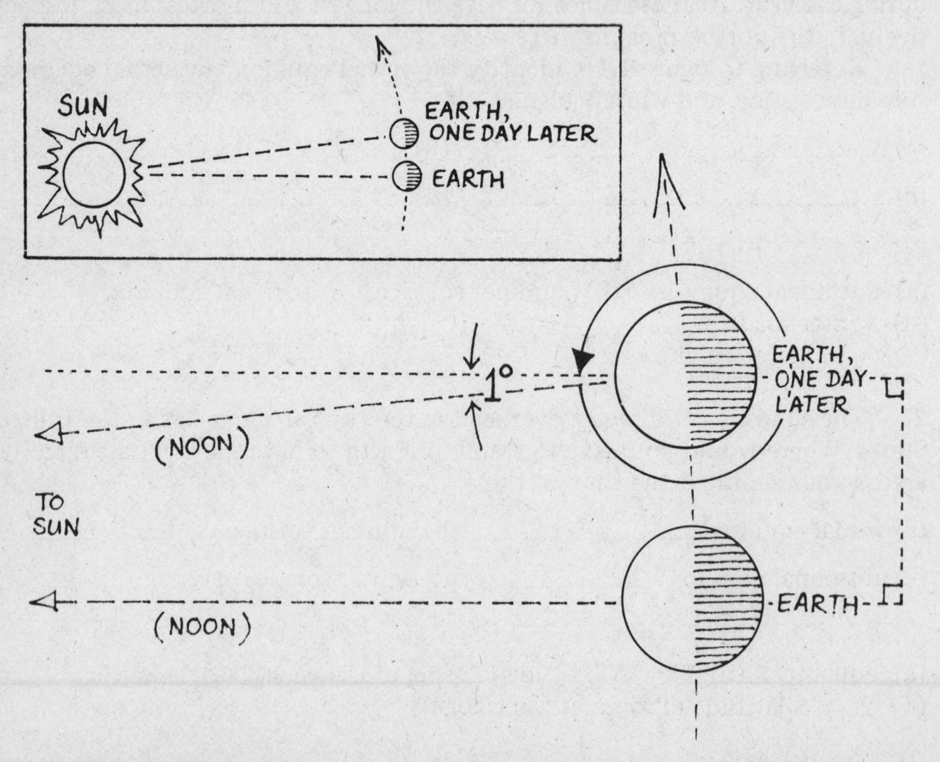

Figure 1.11. Solar (*i*) and sidereal (*ii*) days. Shift in the apparent position of the sun, as Earth moves in orbit.

time interval of Earth's rotation using the stars for reference.

A sidereal day is 23 hours, 56 minutes, 4 seconds long. It is the time interval required for a star to cross your meridian two times successively, or the time for Earth to complete one whole turn in space. A solar day is 24 hours long, the length of time required for two successive meridian transits by the sun (at noon).

A solar day is about 4 minutes longer than a sidereal day because while Earth rotates on its axis it also moves along in its orbit around the sun. Earth must complete slightly more than one whole turn in space before the sun reappears on your meridian.

What motion of Earth causes the 4-minute difference between a sidereal and a solar day?

\- - - - - - - - - - - - - - - - - - -

Earth's revolution around the sun

24. Your star maps will be useful to you for the rest of your life. You may be interested to know, however, that they will finally go out of date thousands of years from now.

The direction of the earth's axis in space shifts extremely slowly around a circle once about every 26,000 years. The slow motion of Earth's axis around a cone in space is called *precession.* The precession of Earth's axis is caused by the tug of the gravity of the sun and moon on Earth's equatorial bulge.

As Earth's axis precesses, the Pole Star changes. The vernal equinox, the zero point of right ascension, drifts westward around the ecliptic at a rate of about 50 seconds per year, or 30° (a whole zodiac constellation) in 2,150 years. Then all star charts go out of date.

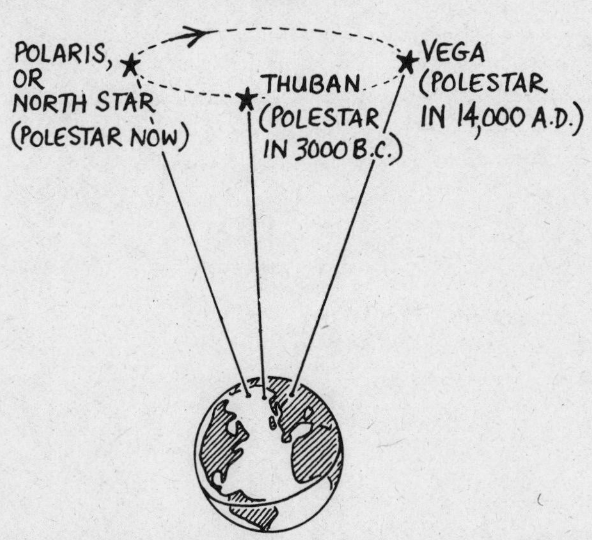

Figure 1.12. The changing Pole Star.

The present Pole Star is Polaris, and the vernal equinox is located in the constellation Pisces. (a) What was the Pole Star in the year 3000 B.C.? ______ ______ (b) What will it be in the year 14,000 A.D.? ______

- - - - - - - - - - - - - - - - -

(a) Thuban; (b) Vega

SELF-TEST

This self-test is designed to show you whether or not you have mastered the material in Chapter One. Answer each question to the best of your ability. Correct answers and review instructions are given at the end of the test.

1. For each of the following references used on a terrestrial globe, list the corresponding name on the celestial sphere:
   (a) equator ________________
   (b) North Pole ________________
   (c) South Pole ________________
   (d) latitude ________________
   (e) longitude ________________
   (f) Greenwich, England ________________

2. Refer to Table 1.1. Which of the five brightest stars in the sky are above the celestial equator, and which are below? ________________
____

3. Refer to Table 1.1. Which of the five brightest stars never appear above the horizon at latitude 40° (about New York City)? ________________
____

4. Match where you might be on Earth with the correct description of the stars:

___ (a) The stars seem to move along circles around the sky parallel to your horizon.

___ (b) The stars rise at right angles to the horizon in the east and set at right angles to the horizon in the west.

___ (c) Vega practically crosses your zenith.

___ (d) Alpha Centauri is always above your horizon.

___ (e) Polaris appears about 30° above your horizon.

(1) Antarctica (below 61°S)
(2) equator
(3) Jacksonville, Florida (30°22′N)
(4) North Pole
(5) Sacramento, California (38°35′N)

5. Why do the stars appear to move along arcs in the sky during the night?

____

6. Why do some different constellations appear in the sky each season?

____

7. What is the zodiac? ______________________________

____

8. Where on Earth would you have to be to have the sun pass directly across your zenith at the time of the (a) vernal equinox? __________ (b) summer solstice?__________ (c) winter solstice?__________

9. If a star rises at 8 P.M. tonight, approximately what time will it rise a month from now? __________

10. Why is a solar day about 4 minutes longer than a sidereal day? ______

____

11. Arrange the following stars in order of decreasing brightness: Antares (magnitude 1); Canopus (magnitude −1); Polaris (magnitude 2); Vega (magnitude 0). ______________________________

12. Why will the Pole Star and the location of the vernal equinox on the celestial sphere be different thousands of years from now, causing your star maps finally to go out of date? ______________________

____

ANSWERS

Compare your answers to the questions on the self-test with the answers given below. If all of your answers are correct, you are ready to go on to the next chapter. If you missed any questions, review the frames indicated in parenthesis following the answer. If you miss several questions, you should probably reread the entire chapter carefully.

1. (a) celestial equator
   (b) north celestial pole
   (c) south celestial pole
   (d) declination
   (e) right ascension
   (f) vernal equinox

   (frames 1, 8, 9)

2. above: Arcturus, Vega below: Sirius, Canopus, Alpha Centauri (frames 9, 10)
3. Canopus, Alpha Centauri (frames 10, 13, 14)
4. (a) 4; (b) 2; (c) 5; (d) 1; (e) 3 (frames 10, 13, 14, 15)
5. because of Earth's rotation (frames 1, 12)
6. because of Earth's revolution around the sun (frame 16)
7. a belt about 16° wide around the sky centered on the ecliptic, containing twelve constellations (frame 17)
8. (a) equator; (b) $23\frac{1}{2}$°N (Tropic of Cancer); (c) $23\frac{1}{2}$°S (Tropic of Capricorn) (frames 19, 20, 21)
9. 6 P.M. (frame 16)
10. Because while Earth rotates on its axis it also moves along in its orbit around the sun. Earth must complete slightly more than one whole turn in space before the sun reappears on your meridian. (frame 23)
11. Canopus, Vega, Antares, Polaris (frame 7)
12. precession of Earth's axis (frame 24)

CHAPTER TWO

Light and Telescopes

> Curiosity is one of the permanent and certain characteristics of a vigorous mind.
>
> Samuel Johnson, *The Rambler* (1751)

1. Most of our information about the universe has been obtained through the analysis of starlight. To explain how starlight travels across billions of miles of empty space to waiting telescopes, astronomers picture light as a form of wave motion.

A *wave* is a rising and falling disturbance that transports energy from a source to a receiver without the actual transfer of material. Wave motion is clearly observable in the ocean; during storms, crashing ocean waves vividly reveal the energy they carry.

A *light wave* is an electromagnetic disturbance. Light waves transport energy from accelerating electric charges in stars (the source) to electric charges in the retina of your eye (the receiver). You become aware of that energy when you see starlight.

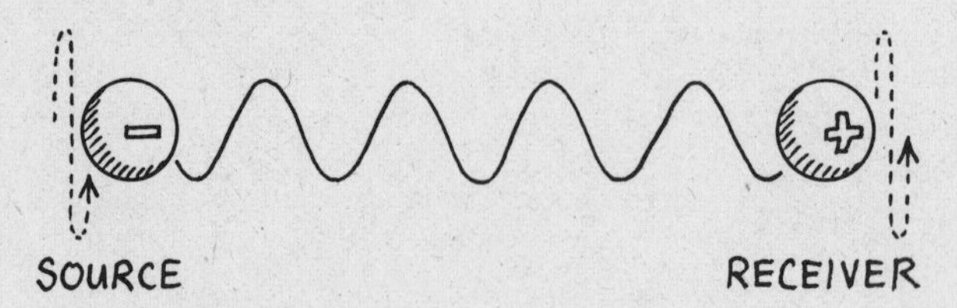

Figure 2.1. Visualizing a light wave.

Explain what a wave is. ____________________

____

A wave is a rising and falling disturbance that transports energy from a source to a receiver without the actual transfer of material.

2. The distance from any point on a wave to the next identical point, such as from crest to crest, is called the *wavelength.*

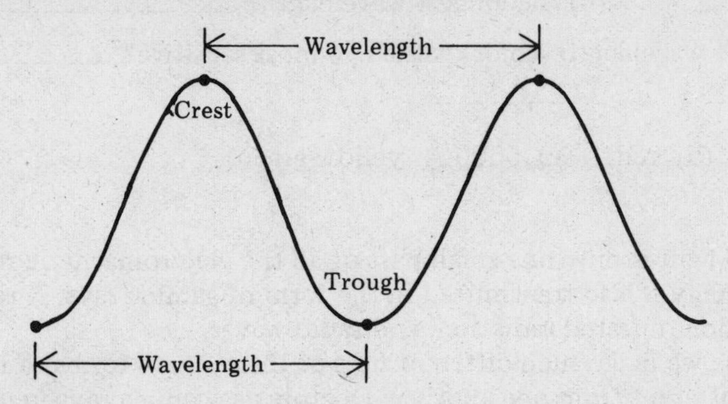

Figure 2.2. Wavelength.

The human eye responds to waves that have extremely short wavelengths. These waves that produce vision are called *visible light.* They are commonly measured in a unit called the *angstrom* (Å). One angstrom equals one hundred millionth of a centimeter (10^{-8}cm). Visible light has wavelengths between 4000 Å and 7000 Å.

The varying wavelengths of visible light are perceived as different *colors.* The arrangement of the colors according to wavelength is called the *visible spectrum.*

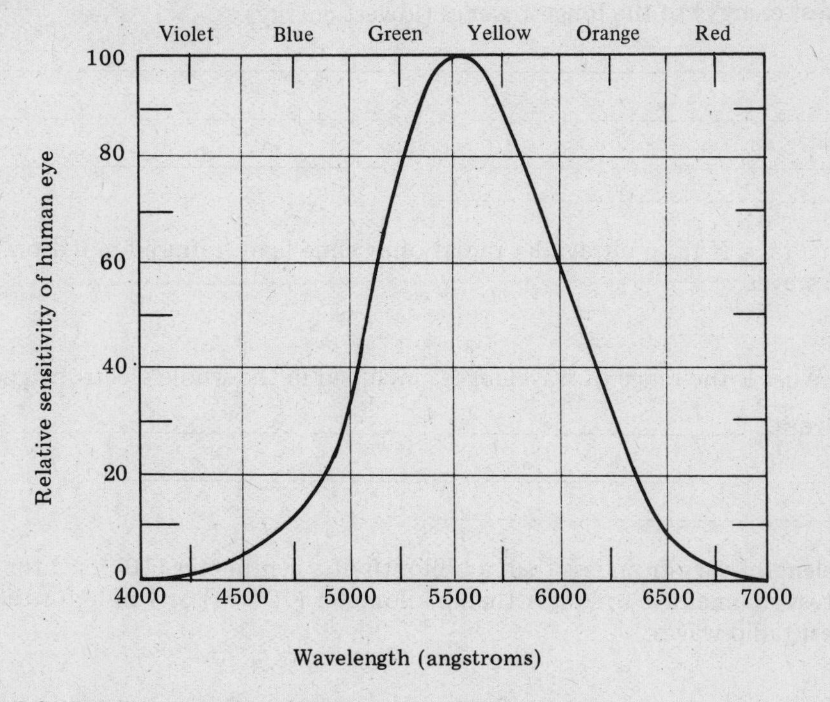

Figure 2.3. Relative sensitivity of human eye to different colors.

Refer to figure 2.3. Which color light has (a) the shortest wavelength?

____________ (b) the longest wavelength? ____________

(c) To which wavelength (color) is the eye most sensitive? ____________

— — — — — — — — — — — — — — — — — —

(a) violet; (b) red; (c) 5550 Å (yellow-green)

3. Visible light is only one small part of all the electromagnetic radiation in space. Energy is also transmitted in the form of gamma rays, X rays, ultraviolet radiation, infrared radiation, and radio waves.

Because we make such different uses of them, these forms of radiation seem very different from one another. Doctors use gamma rays in cancer treatment and X rays for medical diagnosis. Ultraviolet rays give you a suntan, and infrared rays warm you up. Radio waves are used for communication.

However, all of these forms of radiation are really the same basic kind of energy as visible light. They have different properties because they have different wavelengths. The shortest waves have the most energy while the longest waves are the least energetic. The whole family of electromagnetic waves, arranged according to wavelength, is called the *electromagnetic spectrum*, which is illustrated in figure 2.4.

List six forms of electromagnetic radiation from the shortest waves (highest energy) to the longest waves (lowest energy).

____________ ____________

____________ ____________

____________ ____________

— — — — — — — — — — — — — — — — — —

gamma rays, X rays, ultraviolet radiation, visible light, infrared radiation, radio waves

4. What is the range of wavelengths included in the whole electromagnetic spectrum? ____________

____

— — — — — — — — — — — — — — — — — —

Wavelengths vary from less than a billionth of a centimeter (10^{-9} cm) for the shortest gamma rays to longer than a kilometer (10^5 cm) or 1 mile for the longest radio waves.

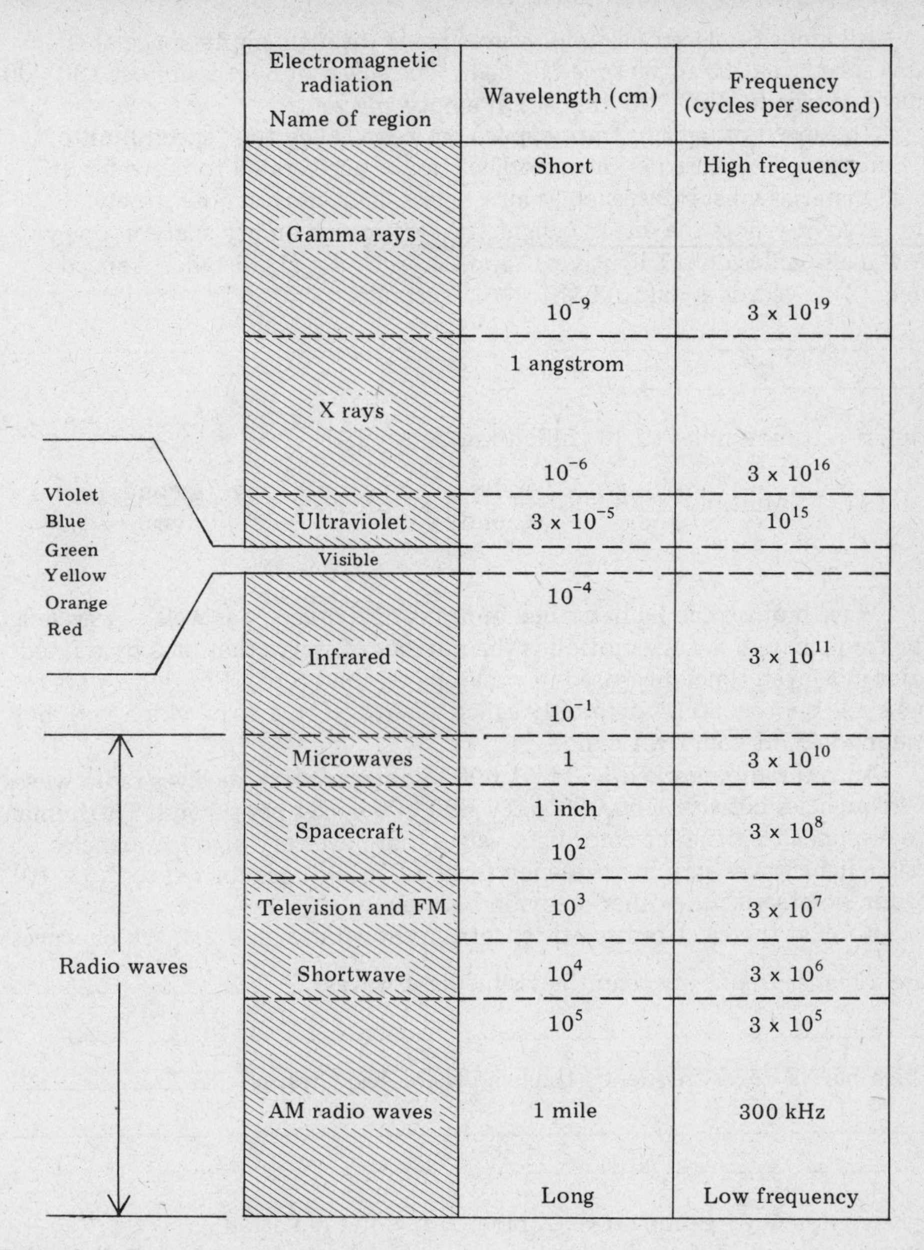

Electromagnetic radiation Name of region	Wavelength (cm)	Frequency (cycles per second)
	Short	High frequency
Gamma rays		
	10^{-9}	3×10^{19}
X rays	1 angstrom	
	10^{-6}	3×10^{16}
Ultraviolet	3×10^{-5}	10^{15}
Visible		
	10^{-4}	
Infrared		3×10^{11}
	10^{-1}	
Microwaves	1	3×10^{10}
Spacecraft	1 inch	3×10^{8}
	10^{2}	
Television and FM	10^{3}	3×10^{7}
Shortwave	10^{4}	3×10^{6}
	10^{5}	3×10^{5}
AM radio waves	1 mile	300 kHz
	Long	Low frequency

Figure 2.4. The electromagnetic spectrum.

5. All kinds of electromagnetic waves move through empty space at the same speed, that is, at the *speed of light.* The speed of light is almost 186,300 miles per second (299,793 km/sec) in empty space.

The speed of light in empty space has been called the "speed limit of the universe," because no known object can be accelerated to move faster. In all material substances, such as air or glass, light moves more slowly.

A *light-year* is the distance light travels through empty space in one year. How many miles does 1 light-year represent? Hints: (1) distance = speed x time. (2) A year is equal to 3.156×10^7 seconds.

Almost 6 trillion miles (9.46 trillion km).

Solution: Multiply $186{,}300\left(\frac{\text{miles}}{\text{second}}\right) \times 3.156 \times 10^7 \left(\frac{\text{seconds}}{\text{year}}\right)$

6. Wave motion can be described in terms of *frequency* as well as wavelength. The frequency of a wave motion is the number of waves that pass by a fixed point in a given time, measured in cycles per second (cps). For radio waves, one cycle per second is commonly called a *hertz* (Hz), a term which you may find marked on your own radio.

An AM radio, marked 550 to 1,600 kilohertz (kHz), receives radio waves of frequencies between 550,000 and 1,600,000 cycles per second. The human eye responds to different color light waves that have very high frequencies. Visible light waves vary in frequency from 4.2×10^{14} cps for red to 7.5×10^{14} cps for violet, with the other colors in between.

Look at the electromagnetic spectrum shown in figure 2.4. Which waves have a higher frequency than the visible light waves? ____________________

Which have a lower frequency than the visible light waves? ____________________

---

higher frequency: gamma rays, X rays, ultraviolet radiation
lower frequency: infrared radiation, radio waves

7. Can you deduce a general relationship between wavelength and frequency for these waves? ____________________

The shorter waves have a relatively higher frequency, and the longer waves have a relatively lower frequency.

8. The relationship you have just found is an example of a formula that holds true for all kinds of wave motion: speed of wave = frequency x wavelength.

You can use this formula to calculate the frequency of any kind of electromagnetic wave in empty space if you know its wavelength (or the wavelength if you know the frequency). Explain why. Hint: Review frame 2.5.

____

____

All electromagnetic waves have the same speed in empty space—that is, the speed of light, or about 186,000 miles per second.

9. Make sure you understand this relationship between speed, frequency, and wavelength for electromagnetic waves. Calculate the wavelength of a radio wave whose frequency is 100 kHz (100,000 cycles per second). __________

____

1.86 miles.

Solution: speed = frequency x wavelength

$$\text{so, wavelength} = \frac{\text{speed}}{\text{frequency}} = \frac{186{,}000 \text{ miles/sec}}{100{,}000 \text{ cycles/sec}}$$

10. ★ Stars, like other hot bodies, radiate energy of all different wavelengths. The hotter the star, the more radiant energy it emits. The temperature of a star determines which wavelength is brightest.

Wien's law of radiation states that the wavelength at which a body emits the greatest amount of radiation is inversely proportional to its temperature. Thus the hotter a star is, the shorter the wavelength at which it emits its maximum light.

Some stars are thousands of degrees hotter than others. You can judge how hot a star is by its color (wavelength). The hottest stars look blue-white (short wavelength), and the coolest stars look red (long wavelength). Extremely hot stars (very short wavelengths) and extremely cool stars (very long wavelengths) are not visible.

Look in the sky for the examples cited in Table 2.1 (see star maps):

Table 2.1 Hot and cool stars

Season	Star	Constellation	Color	Surface Temperature (°K)
Summer	Vega	Lyra	Blue-white	10,000
Summer	Antares	Scorpio	Red	3,000
Winter	Sirius	Canis Major	Blue-white	10,000
Winter	Betelgeuse	Orion	Red	3,400

A radiation curve shows how much energy a body radiates at different wavelengths, which wavelengths it radiates most intensely, and the total amount of energy it radiates at all wavelengths (indicated by the area under the curve).

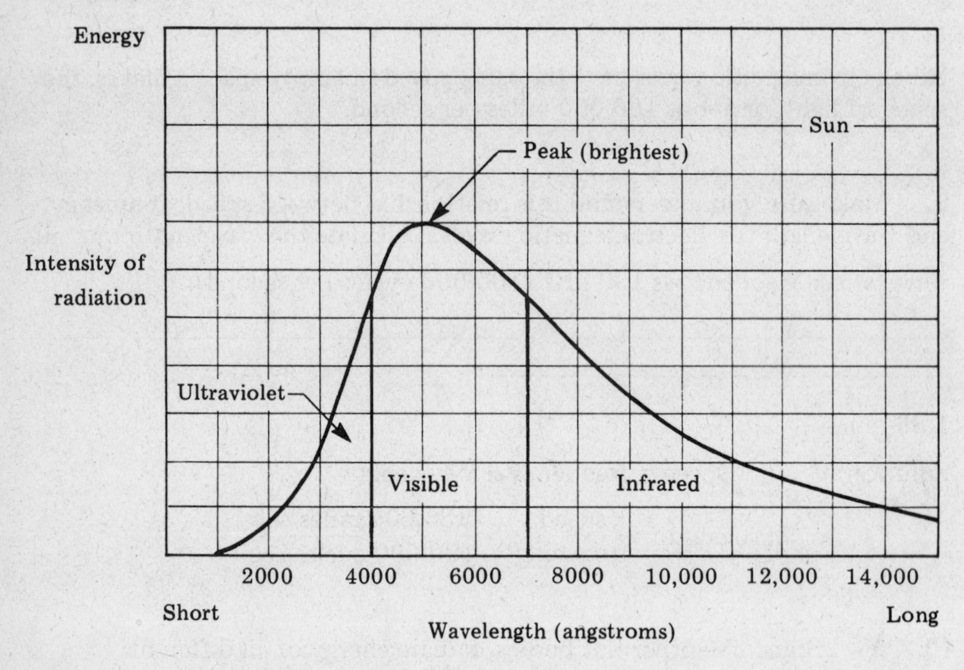

Figure 2.5. The sun's radiation curve.

Examine the radiation curve for our sun. (a) The sun radiates most intensely in the ________________ wavelengths. (b) The total amount of energy that the sun radiates as visible light is (more, less) __________ than the amount radiated outside the visible region.

— — — — — — — — — — — — — — — — — — —

(a) visible; (b) less

11. Electromagnetic waves of all wavelengths are important to astronomers, because all such waves carry information about the universe. Earth's atmosphere stops most radiation from space and permits only certain wavelengths to shine through to telescopes on the ground.

Ground-based astronomers can look out at the universe through three windows, or wavelength bands. These windows are, in order of their importance, the *optical* (visible), *radio*, and *infrared* bands. Earth's atmosphere is transparent to waves in these three wavelength bands.

Figure 2.6. Kitt Peak National Observatory

The site of the Kitt Peak National Observatory, shown in figure 2.6, was carefully chosen for optical astronomy observations because of the frequently clear skies in the mountains of the arid Southwest. The Observatory, about 50 miles outside of Tucson, Arizona, has a visitor's center and an interesting self-guided tour for the public.

What would you suggest to astronomers who want to observe the universe using gamma rays, X rays, and ultraviolet radiation?

- - - - - - - - - - - - - - - - - -

Get their instruments up above Earth's atmosphere. Today space age technology makes observations in these wavelength bands possible from balloons, rockets, Earth-orbiting satellites, or even from moon-based observing stations.

12. *Optical telescopes*, which form images of faint and distant stars, can collect much more light from space than the human eye can. Optical telescopes are built in two basic designs—*refractors* and *reflectors.*

The heart of any telescope is its *objective.* The objective is a lens (in refractors) or mirror (in reflectors) whose function is to gather light from a sky object and form its image. The amount of light collected by the objective is called its *light-gathering power.* Light-gathering power is proportional to the area, or to the square of the diameter, of the collector. The *size* of a telescope, such as "6-inch" or "200-inch," refers to the diameter of its objective.

You can look at the image formed by the objective through an *eyepiece*, which is a magnifying glass. Or you can photograph the image or analyze it in a variety of other ways. Your eye lens size is about $\frac{1}{5}$ inch. A 6-inch telescope has an objective that is thirty times bigger than your eye lens. Its light-gathering power is $(30)^2$, or nine hundred times greater than your eye's. So a star appears nine hundred times brighter with a 6-inch telescope than it does to your naked eye.

All stars appear brighter with telescopes than they do to your eye alone. All the extra starlight gathered by the telescope is concentrated into a single point. Using time exposure, the 200-inch telescope can photograph very faint stars down to nearly magnitude 24 (see frame 1.7). A star of that magnitude has about the same apparent brightness as a candle viewed from 10,000 miles away!

How much brighter will a star appear with the 200-inch (508-cm) telescope than to your naked eye? Explain. ______________________________

____

____

— — — — — — — — — — — — — — — — — — — —

Over one million times brighter. The 200-inch telescope is over one thousand times bigger than your eye lens, so it gathers over $(1{,}000)^2 = 1$ million times more light.

13. Binoculars labelled 7X50 have an objective 50 millimeters in diameter. The 7X specifies magnification, discussed in frame 2.20.

Why do binoculars and telescopes reveal many more sky objects than you can see with your naked eye? ______________________________

____

— — — — — — — — — — — — — — — — — — — —

They can collect much more light than your eye can. (Light-gathering power is proportional to the area of the objective.)

14. The *refracting telescope* has a large lens (the objective) permanently mounted at the front end of the tube. Starlight enters this lens and is *refracted*, or bent, so that it forms an image near the back of the tube.

The distance from this lens to the image is its *focal length.* You look at the image through a removable magnifying lens called the *ocular*, or eyepiece. The tube keeps out scattered light, dust, and moisture.

Galileo Galilei first pointed refracting telescopes skyward in Italy in the seventeenth century. The largest instrument he made was smaller than 2 inches. Today refracting telescopes range in size from a beginner's 2.4-inch (60-mm) version to the world's largest, the 40-inch (102-cm) telescope at the Yerkes Observatory in Williams Bay, Wisconsin.

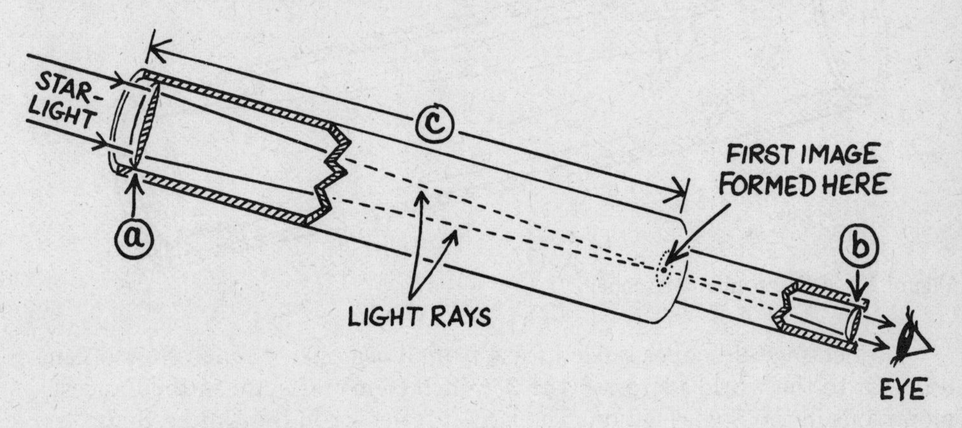

Figure 2.7. A refracting telescope.

Referring to figure 2.7, identify the refracting telescope's objective lens, the ocular, and the focal length of the objective. State the purpose of (a) and (b).

(a) ____________________

(b) ____________________

(c) ____________________

— — — — — — — — — — — — — — — — — — — —

(a) objective lens: to gather light and form an image
(b) ocular: to magnify the image formed by the objective
(c) focal length of objective

15. The *reflecting telescope* has a highly polished curved-glass mirror (the objective) mounted at the bottom of an open tube. When starlight shines on this mirror, it is reflected back up the tube to form an image at the *prime focus.*

You can place photographic film at the prime focus to record the image, or you can use additional mirrors to reflect the light to another spot for viewing. The *Newtonian reflector* uses a small flat mirror to reflect the light through the side of the tube to an eyepiece. The *Cassegrain reflector* uses a small convex mirror to reflect the light through a hole cut in the objective mirror at the bottom end of the tube.

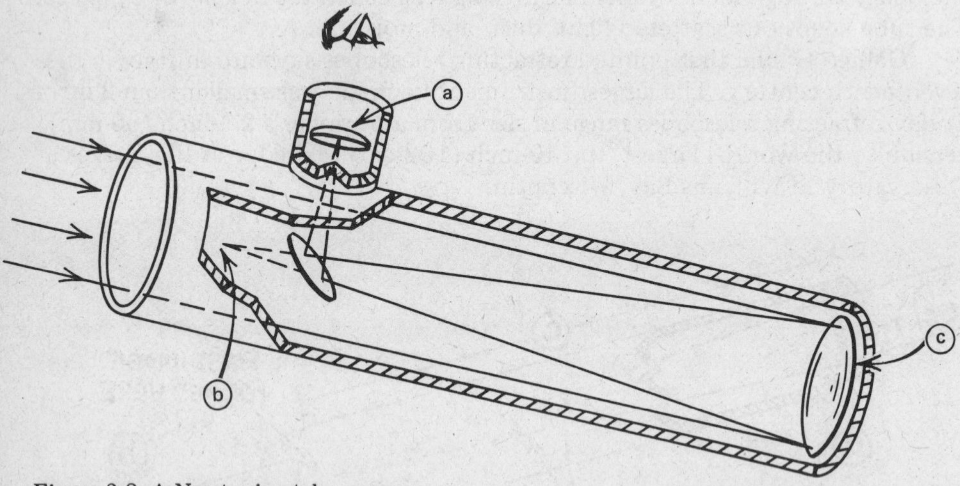

Figure 2.8 A Newtonian telescope.

Reflecting telescopes range in size from a beginner's 3-inch Newtonian reflector to the world's largest—the 236-inch (6-m) reflector in the Caucasus mountains in the Soviet Union. The largest reflector in the United States is the 200-inch (508-cm) Hale Telescope on Mount Palomar in California.

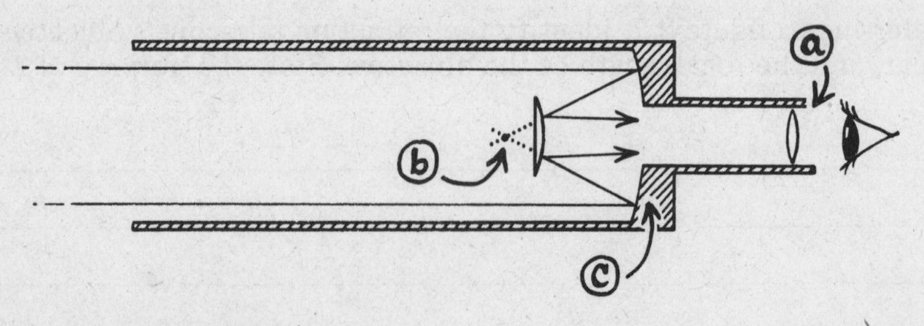

Figure 2.9 A Cassegrain reflecting telescope.

Identify the reflecting telescope's objective mirror, ocular, and prime focus in figure 2.8 and figure 2.9.

(a) ______________; (b) ______________; (c) ______________;

- -

(a) ocular; (b) prime focus; (c) objective mirror

16. What is the essential difference between a reflecting and a refracting telescope? Explain. ____________________

____

– – – – – – – – – – – – – – – – – – – –

The type of objective. A reflecting telescope uses a mirror, while a refracting telescope uses a lens.

17. Telescopes are often described by both their objective size and *f number.* The *f* number is the ratio of the focal length of the objective to its diameter. These specifications are important because the brightness, size, and clarity of the image produced by a telescope depend on the diameter and focal length of its objective. For example, a "6-inch, *f*/8 reflector" means the objective mirror is 6 inches in diameter and has a focal length of 48 inches (8 times 6).

What is the focal length of the 200-inch, *f*/3.3 mirror on Mount Palomar?

____

– – – – – – – – – – – – – – – – – – – –

660 inches, or 55 feet

18. Image *size* is proportional to the focal length of the telescope's objective. For example, a mirror whose focal length is 100 inches produces an image of the moon that measures almost 1 inch across. You know that the 200-inch mirror has a focal length of 660 inches, which is over six times longer. Hence, it produces an image of the moon that is about six times bigger, or about 6 inches across.

Lenses and mirrors form real images that are upside down. (A real image is formed by the actual convergence of light rays.) All stars except our sun are so far away that they appear as dots of light. The moon and planets appear as small disks. Since inverted images do not matter in astronomical work, nothing is done to turn them upright in telescopes.

What determines the size of the image formed by a telescope? ________

____

– – – – – – – – – – – – – – – – – – – –

the focal length of the objective

19. Even if a telescope were of perfect optical quality, it would not produce perfectly focussed images because of the nature of light itself. A telescope's *resolving power* is its ability to produce sharp, detailed images under ideal observing conditions. Resolving power is proportional to the diameter of the objective.

Starlight travels in straight lines through empty space. But when waves of starlight pass close to the edge of a lens or mirror, they spread out, in an effect called *diffraction*, and come to a focus at different spots. Because of diffraction, the image of a star formed by a lens or mirror appears as a tiny blurred disk surrounded by faint rings, called a diffraction pattern, instead of as a single point of light.

Figure 2.10. Diffraction pattern.

If two stars are close together, their diffraction patterns may overlap so that they look like a single star. Features such as moon craters and planet markings are also blurred by diffraction.

Resolving power indicates the smallest angle between two stars for which separate, recognizable images are produced. The resolving power of the human eye is about one minute of arc ($1'$). Explain why what may look like a single star to the naked eye may resolve into two close neighbor stars in a telescope.

- - - - - - - - - - - - - - - - - -

Resolving power is proportional to objective diameter, and a telescope has a much larger objective diameter than the diameter of the naked eye.

20. A telescope's *magnifying power* is the ratio of the apparent size of an object seen through the telescope compared to its size when seen by the naked eye. Telescopes magnify the angular diameter of objects. Thus the image appears to be closer than the object.

For example, to your naked eye the apparent size of the full moon is $\frac{1}{2}^\circ$, the same as an aspirin held at arm's length. If the apparent size of the moon increases twenty times (so that it looks 10° in diameter) when you view it through your telescope, then the magnifying power is 20, written 20X.

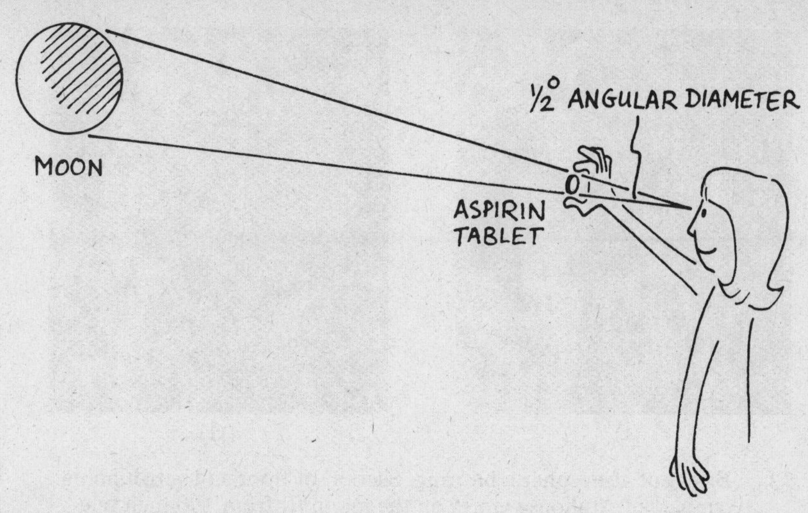

Figure 2.11. Angular diameter.

The value of the magnifying power of a telescope is:

$$\text{magnifying power} = \frac{\text{focal length of objective}}{\text{focal length of eyepiece}}$$

A telescope usually comes with several eyepieces of different focal lengths so you can vary its magnifying power. (a) What is the magnifying power of a 6-inch, $f/8$ telescope when an eyepiece of $\frac{1}{2}$-inch focal length is used? ______________________________

(b) How could you increase the magnifying power of this telescope?

____

(a) 96X.

Solution: $\text{magnifying power} = \frac{\text{focal length of objective}}{\text{focal length of eyepiece}} = \frac{48 \text{ inches}}{\frac{1}{2} \text{ inch}}$

(b) Use an eyepiece of shorter focal length.

21. It is a mistake to exaggerate the importance of magnifying power when you buy a telescope. You cannot increase the useful magnifying power indefinitely by changing eyepieces.

Starlight must pass through Earth's atmosphere to reach waiting telescopes on the ground. Disturbances in the air cause blurry images. *Seeing* refers to atmospheric conditions which affect how sharp a telescope image is. If the air is quiet, then the "seeing" is good, and stars shine with a steady light. If the air is turbulent, then the "seeing" is bad, and the stars twinkle madly.

The *practical limit of useful magnification* for any telescope is about fifty times its objective diameter in inches. Higher power will serve to magnify

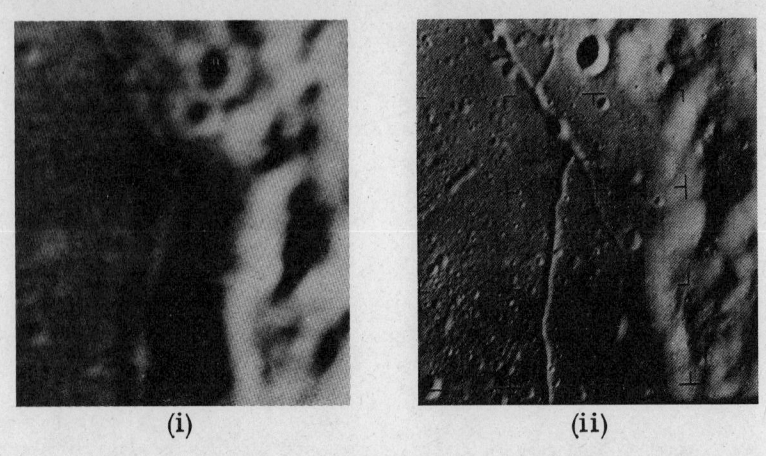

Figure 2.12. Effect of atmospheric blurring. Section of floor and surrounding rampart of Alphonse crater on the moon (*i*) from 100-inch telescope, Mt. Wilson; (*ii*) from spacecraft (above atmosphere).

any blurring in the image due to diffraction or bad seeing. It can't reveal any finer details.

Space Telescope is NASA's 10-ton observatory that is designed to orbit a 94-inch telescope above the atmosphere in the 1980's. Astronomers will operate it by remote control from the ground. Astronauts will maintain and change its instruments in orbit. They will return it to Earth for extensive overhaul.

What is the practical limit of useful magnification for a 6-inch telescope?

300X

22. Image *aberrations* are imperfections produced by the objective.

Chromatic aberration is a lens defect. Starlight consists of all the colors of the spectrum. When starlight passes through a lens, the lens bends blue (shorter) light waves most and brings them to a focus closer to the lens than red (longer) light waves. This variation blurs the star image with color fringes. An *achromatic lens*, a combination of two different lenses made of different kinds of glass, reduces this defect.

A mirror reflects all the colors of starlight to a focus at the same point. The image formed by a reflector has no color fringing.

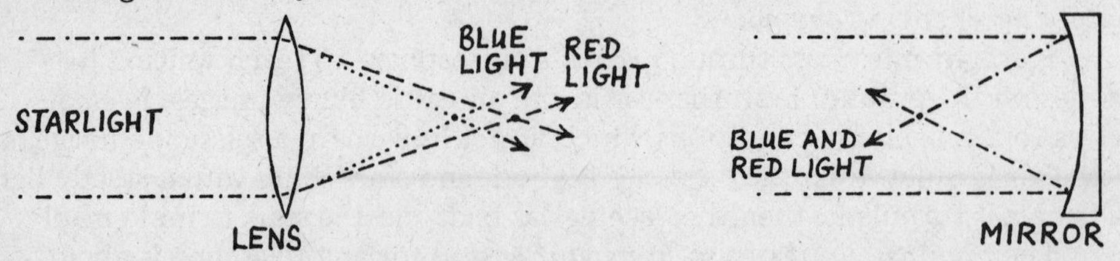

Figure 2.13. Light focussing by lens (*i*) and mirror (*ii*).

Spherical aberration is a defect of spherical lenses and mirrors—the most common variety. Starlight passing close to the edge of the lens or mirror does not focus at the same spot as light going through the center. This defect is corrected by making lenses out of a combination of different kinds of glass and by making parabolic mirrors.

Which type of telescope—reflector or refractor—would you use to photograph stars in color to avoid chromatic aberration? Explain your answer.

- - - - - - - - - - - - - - - - - - - -

Reflector. Because the mirror reflects all the colors of starlight to a focus at the same point, the image formed has no color fringing.

23. You probably wonder which type of telescope is better—a refractor or a reflector. The answer depends on the application involved since each type has advantages and disadvantages when compared to the other.

Small telescopes for astronomy enthusiasts can be of either design. Refractors require less maintenance. But reflectors are cheaper per inch of aperture and are easier to make. The 6-inch Newtonian reflector is very popular with amateurs.

Large refractors are used where image quality and resolution are most important, as for viewing surface details of planets or observing *binary stars* (two stars revolving around each other).

Giant reflectors are used to probe the faintest, most distant objects in the observable universe. Reflectors can be designed using *folded optics*, thus reducing their length so they can be housed inside smaller domes than refractors. The primary mirror can be supported from behind so that it doesn't sag under gravity as lenses do.

Huge reflectors are cheaper and easier to build and are more cost-effective in performance than refractors. Astronomers are designing ever larger telescopes. The *Multiple Mirror Telescope* (MMT) is an innovative prototype for visible and infrared observations. The MMT synchronizes light collection by six 72-inch mirrors to perform like a single 176-inch mirror. A single image is maintained by an electronic control system using smaller movable mirrors, lasers, and computers.

Table 2.2. The five largest reflector telescopes in the world.

Name of Observatory	Location	Size of Mirror (inches)	(meters)
1. Special Astrophysical	Mount Pastukhov, U.S.S.R.	236	6
2. Palomar	Palomar Mountain, California, U.S.A.	200	5
3. Smithsonian Astrophysical	Mount Hopkins, Arizona, U.S.A.	176*	4.5
4. Inter-American	Cerro Tololo, Chile	158	4
5. Kitt Peak National	Near Tucson, Arizona, U.S.A.	158	4

*MMT (see frame 2.23)

(a) Where is the largest optical telescope in the southern hemisphere?

____

(b) How large is it? ______________________________

(a) Inter-American Observatory, Cerro Tololo, Chile. (b) 158 inches

24. With research time in great demand, no single astronomer can sit at the giant telescopes and simply stargaze. Instead, starlight (directly or after passing through electronic imaging systems) is recorded on photographic film or in a data bank for later exhaustive study by many scientists.

Often an instrument called a *spectrograph* is attached to the telescope. Starlight is not a single color but rather a mixture of colors, or wavelengths. Astronomers deduce much information about stars from these separate wavelengths, as you will see in Chapter Three.

A spectrograph separates starlight into its component wavelengths. Starlight enters the spectrograph through a narrow slit and goes through a *collimating lens*, which produces a beam of parallel rays of light. A prism or grating disperses this light into its separate colors (wavelengths). A camera records this spectrum on a photographic plate.

You can produce a spectrum from sunlight (starlight). Place a mirror in a pan of water so that it is under the water and leaning against the side of the pan. Position the pan in bright sunlight so that the sun shines on the mirror. Move the mirror slightly until you see a spectrum on the ceiling or wall.

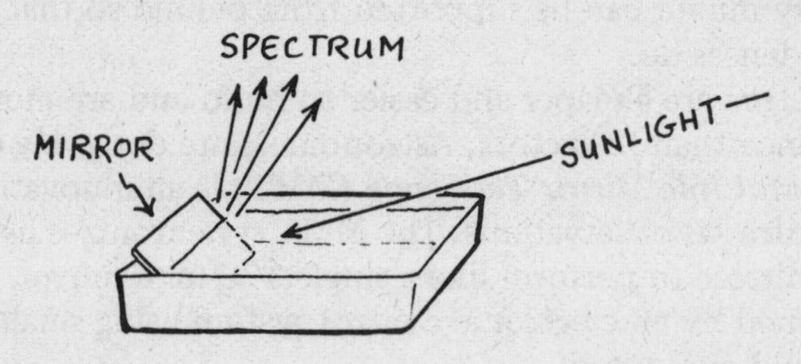

Figure 2.14. Producing a spectrum.

What is the purpose of a spectrograph? ______________________________

____

---

to separate and photograph the individual wavelengths in a beam of light

25. New kinds of telescopes let today's astronomers "look" farther into space and "see" more fascinating sights than at any time in the past.

Radio telescopes reveal information about the universe that optical telescopes could never show. Radio telescopes use a huge curved "dish" to collect and focus *radio waves* from space.

You cannot see or hear or photograph these radio waves directly. Instead, an antenna detects the radio waves, and a tuned radio receiver amplifies and records their electronic image. Programmed computers translate these electronic signals into a *radiograph*, which is an optical picture that shows how the radio source in space would "look" to a person with "radio vision."

Radio astronomy was born in 1931 when K. G. Jansky discovered radio waves coming from the Milky Way. In the last 20 years radio waves have been received from diverse sources including our sun, some planets, galaxies, and the puzzling celestial objects known as quasars.

Radio telescopes must be very big to collect long radio waves and produce clear images. The world's largest dish is the 1,000-foot collector built into a natural bowl in the hills near Arecibo, Puerto Rico. The largest steerable radio telescope in the United States is the 300-foot (90-m) one at the National Radio Astronomy Observatory in Greenbank, West Virginia.

Figure 2.15. A radio telescope.

Identify the dish and the antenna on a radio telescope shown in figure 2.15, and explain the purpose of each. (a) ______________________

____

____

(b)____

- - - - - - - - - - - - - - - - - - -

(a) antenna: to detect these radio waves
(b) dish: to collect radio waves from space

26. Radio telescopes have several advantages. They let us "see" many celestial objects that emit powerful radio waves but little visible light. They let us "see" radio sources behind interstellar dust clouds in our Milky Way Galaxy that blot out visible stars (because radio waves pass through these clouds). Our atmosphere does not stop or scatter radio waves, so radio telescopes can be used in cloudy weather and during the daytime. They have much larger collecting dishes than reflectors, so they can gather more radiation and extend our "vision" much farther into the mysterious darkness of space.

As with optical telescopes, still more and clearer radio data can be gleaned by ever larger collectors. *Aperture synthesis* combines observations from two or more radio telescopes linked electronically with computers to imitate the performance of a single giant collecting dish.

The world's most powerful radio astronomy tool is the *Very Large Array* (VLA), the primary facility of the National Radio Astronomy Observatory. It can closely duplicate the performance of a fully steerable radio dish 17 miles (27 km) in diameter. Located in central New Mexico, the VLA has 27 movable 82-foot radio dishes. These are deployed at 72 observing stations to operate as 351 two-antenna combinations called *interferometers*. A complex computer controls the antennas, processes and displays observed data, and produces radiographs with resolution equal to giant optical reflector photographs.

A technique called *Very Long Baseline Interferometry* (VLBI) combines observations from two or more radio telescopes linked electronically at great distances apart to reveal information about radio sources.

The National Aeronautics and Space Administration (NASA) has radio telescopes on three continents in the *Deep Space Network*. Stations in California, Spain, and Australia are used for VLBI observations, as well as for space missions. Each has receiving, transmitting, data handling, and interstation communication equipment. The control center is located at the Jet Propulsion Laboratory in California.

List at least three advantages of a radio telescope. ________________

____

- - - - - - - - - - - - - - - - - - -

1. reveals radio sources—objects that shine in the radio band of wavelengths
2. shows radio sources behind interstellar dust clouds in parts of the Milky Way that are hidden from optical viewing
3. works in cloudy weather and daytime
4. shows radio sources that are located beyond our power of optical viewing

27. *Infrared telescopes* are basically optical reflectors with a special heat detector at the prime focus. Detectors are shielded and cooled to about 4° K to ensure that they register *infrared rays* from space, rather than stray heat from people, equipment, and observatory walls.

Water vapor in the air strongly absorbs incoming infrared rays. Frank Low built a sensitive infrared detector for astronomy in 1963. Today large infrared telescopes are located on very high mountain tops where the air over head is thinnest and driest. The largest is the United Kingdom's 150-inch (3.8 m) reflector on Mauna Kea's 13,800-foot summit in Hawaii. Smaller telescopes are lofted in planes, balloons, rockets, and spacecraft.

Infrared sources such as solar system objects, newly forming stars, and interstellar clouds are relatively cool and may not be visible (frame 2.10). Infrared rays pass through interstellar dust more readily than shorter visible rays, revealing the nature of different parts of our Galaxy. They are not blotted out by sunshine so infrared telescopes can work day and night.

Why are infrared telescopes located on very high, dry sites? __________

____

Elsewhere the air's water vapor absorbs most incoming infrared rays.

28. *High energy astrophysics* is a young, dynamic field where most discoveries have come during the last decade. *Ultraviolet*, *X-ray*, and *gamma ray telescopes* with suitable detectors are sent above the Earth's obscuring air in orbiting spacecraft. They collect radiation and transmit data to the ground for analysis.

Computers commonly process the data. They may visually display processed data, store it, or enhance input and generate spectacular false color pictures.

New observations of the sun, hot stars, stellar atmospheres, interstellar clouds, a hot gas galactic halo, and extragalactic sources abound from the International Ultraviolet Explorer (IUE) launched in 1978.

Myriad X-ray and gamma ray sources were observed by three unmanned orbiting High Energy Astronomy Observatories (HEAO) launched between 1977–79. HEAO effectively extended our X-ray "vision" as far into space as the best optical and radio telescopes observe. Fascinating data on cosmic bursters (sudden intense bursts of radiation), pulsars, possible black holes, active galaxies, and distant quasars promise new insights into the universe.

What is particularly interesting about new observations by ultraviolet, X-, and gamma ray telescopes? Hint: Review frames 2.3 and 2.10.

____

Incoming ultraviolet, X-, and gamma rays have much more energy than visible light. They must be generated in extraordinarily energetic processes not yet comprehended.

SELF-TEST

This self-test is designed to show you whether or not you have mastered the material in Chapter Two. Answer each question to the best of your ability. Correct answers and review instructions are given at the end of the test.

1. Explain why looking at stars is a way of seeing how the universe looked many years ago. ______________________________

   ____

2. (a) List the major regions of the electromagnetic spectrum from shortest wavelength (highest energy) to longest wavelength (lowest energy).

   ____

   (b) State what all electromagnetic waves have in common. ______________

   ____

3. Write the general formula which relates the wavelength and frequency of a wave. ______________________________

4. Suppose you observe a bluish star and a reddish star in the sky. State which is hotter, and explain how you know. ______________

   ____

5. List the three windows (wavelength bands) in Earth's atmosphere in their order of importance to observational astronomy. ______________

   ____

6. What are the two main parts of a telescope used for stargazing, and what is the function of each? ______________________________

   ____

7. Which telescope described on the following page (1 or 2) has:

   ___ (a) greater light-gathering power?

   ___ (b) greater resolving power?

   ___ (c) greater magnification?

Characteristic	Type of Telescope	
	Reflector (1)	Refractor (2)
Diameter of objective lens or mirror	72 inches	36 inches
Focal length of objective	300 inches	48 feet
Focal length of eyepiece	2 inches	$\frac{1}{2}$ inch

8. What two factors are most important in telescope performance?
____

9. What is the purpose of a spectrograph? ______________________________
____

10. List three advantages of a radio telescope. ______________________________
____

11. What is the advantage of sending telescopes up in spacecraft?
____

12. Match an appropriate innovative tool to the observations.

___ (a) faintest and most distant radio sources
___ (b) hot stars and gas
___ (c) visible and relatively cool celestial sources
___ (d) X-ray and gamma ray sources

1. High Energy Astronomy Observatory (HEAO)
2. International Ultraviolet Explorer (IUE)
3. Multiple Mirror Telescope (MMT)
4. Very Large Array (VLA)

ANSWERS

Compare your answers to the questions on the self-test with the answers given below. If all of your answers are correct, you are ready to go on to the next chapter. If you missed any questions, review the frames indicated in parentheses following the answer. If you miss several questions, you should probably reread the entire chapter carefully.

1. Starlight is radiated by electric charges in stars. Light waves transport energy from stars to electric charges in our eyes. Light waves travel incredibly fast—about 186,000 miles per second. But trillions of miles separate the stars from Earth, and the journey takes many years. Thus we see the stars as they were many years ago when the starlight began its journey to Earth.
(frames 1, 5, 10)

2. (a) gamma rays, X rays, ultraviolet radiation, visible light, infrared radiation, radio waves; (b) All electromagnetic waves travel through empty space at the same speed, the speed of light—about 186,000 miles per second.

   (frames 3, 5, 10)

3. $$\text{wavelength} = \frac{\text{speed of wave}}{\text{frequency}}$$ (frames 2, 6, 7, 8)

4. The bluish star is hotter. The shorter the wavelength at which a star emits its maximum light, the hotter the star is, according to Wien's law of radiation. Blue light has a shorter wavelength than red light.

   (frames 2, 10)

5. optical, radio, infrared (frame 11)

6. (1) objective: to gather light and form an image
   (2) eyepiece: to magnify the image formed by the objective

   (frames 12, 13, 14, 15)

7. (a) 1; (b) 1; (c) 2 (frames 12, 19, 20)

8. objective size and quality (frames 12, 17, 18, 19, 20, 21, 22)

9. to separate and photograph the individual wavelengths in a beam of light

   (frame 24)

10. reveals radio sources, shows radio sources behind interstellar dust clouds in parts of the Milky Way that are hidden from optical viewing, works in cloudy weather and daytime, shows radio sources that are located beyond our power of optical viewing (frames 25, 26)

11. Spacecraft take the telescopes above Earth's obscuring atmosphere, where it is possible to observe gamma rays, X rays, and ultraviolet sources that cannot be observed on the ground. There is no atmospheric blurring or radio interference, so a space telescope can work at its practical limit of resolving power. (frames 11, 19, 21, 25, 27, 28)

12. (a) 4; (b) 2; (c) 3; (d) 1 (frames 23, 26, 27, 28)

CHAPTER THREE

The Stars

Look at the stars! look, look up at the skies!
O look at all the fire-folk sitting in the air!
Gerard Manley Hopkins,
"The Starlight Night"

1. The huge fiery stars are really trillions of miles above our atmosphere. The difficult problem of ascertaining the vast distances to the stars has challenged astronomers for centuries.

The method of *parallax* is used in measuring the distances to nearby stars. The position of a star is carefully determined relative to other stars. Six months later, when Earth's revolution has carried telescopes halfway around the sun, the star's position is measured again.

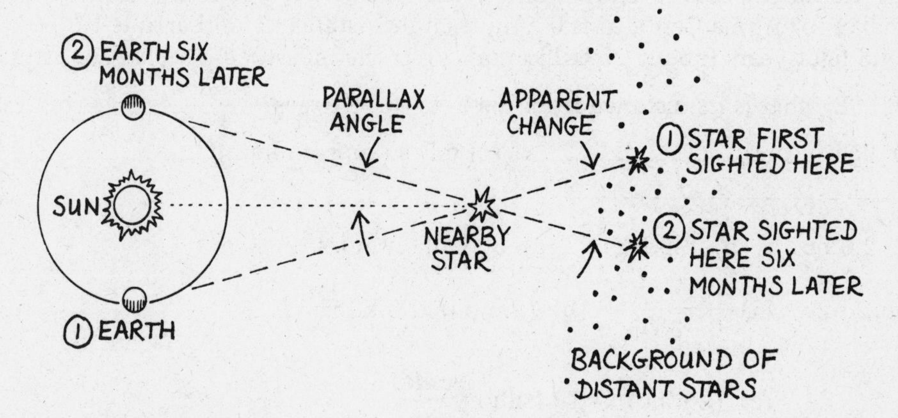

Figure 3.1. Stellar parallax.

Nearby stars appear to shift back and forth among the more distant stars as Earth revolves around the sun. The apparent change in a star's position observed when the star is sighted from opposite sides of Earth's orbit is called *stellar parallax.*

The distance to the star is calculated from its *parallax angle*, one half of the apparent change in its angular position. Stellar parallaxes are very small and are measured in seconds of arc, where $1'' = \frac{1}{3600}^\circ$. One *parsec* (pc) is the distance to a star that has a *par*allax of one *sec*ond of arc (1″). One parsec equals about 19 trillion miles, or 3.26 light-years (LY).

To calculate the distance to any star from its measured parallax, use the formula:

$$\text{star's distance (in pc)} = \frac{1}{\text{parallax ('')}}$$

Stellar parallax decreases with the distance of a star. Stellar parallaxes can be measured down to about one hundredth of a second (0″.01) corresponding to a distance of 100 parsecs (326 LY). However, only the tiny fraction of stars within a distance of about 20 parsecs have parallaxes large enough to be measured with a precision of 10 percent or better. Other indirect methods must be used to determine the distance to the great majority of stars beyond that region.

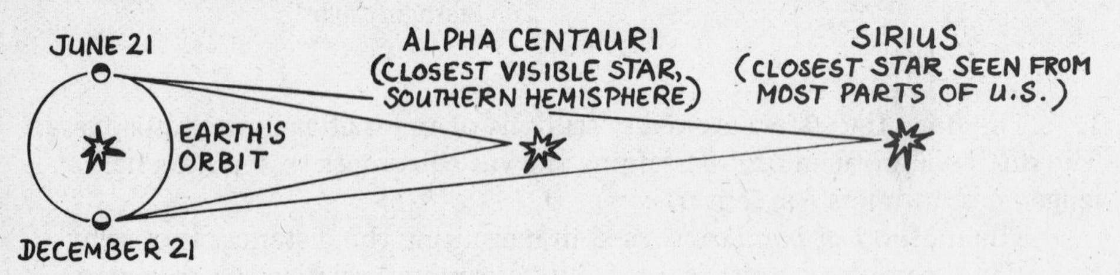

Figure 3.2. Using parallax method.

Would you like to know what "close" means for a star? If the measured parallax for Alpha Centauri is 0″.76, then its distance from Earth is 1.3 pc, or 4.3 light-years (about 25 trillion miles). If the measured parallax for Sirius is 0″.38, what is its distance from Earth in (a) parsecs? ____________

(b) light-years? ____________ (c) miles (approximate)? ____________

— — — — — — — — — — — — — — — — —

(a) 2.6 pc; (b) 8.5 LY; (c) 50 trillion miles

Solution: (a) $\frac{1}{0''.38}$; (b) $(2.6 \text{ pc}) \times 3.26 \frac{\text{LY}}{\text{pc}}$;

(c) $(2.6 \text{ pc}) \times 19 \text{ trillion} \frac{\text{miles}}{\text{pc}}$

2. Despite the vast distances that separate us from the stars, we know a lot about them. Astronomers can extract an amazing amount of information from starlight.

Remember that starlight is composed of many different wavelengths. When starlight is separated into its component wavelengths, the resulting spectrum holds many clues about the stars. *Spectroscopy* is the analysis of spectra (or spectrums). Spectra are of three different types, as illustrated in figure 3.3.

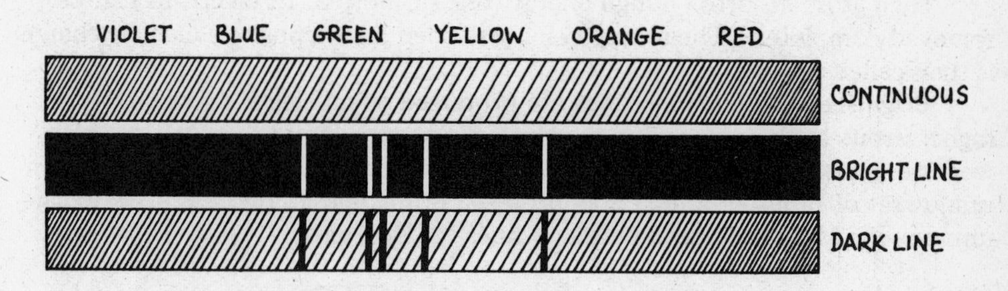

Figure 3.3. Three types of spectra.

Describe the appearance of each type.

(a) ____________________

____

(b) ____________________

____

(c) ____________________

____

- - - - - - - - - - - - - - - - - -

(a) *continuous spectrum:* a continuous array of all the rainbow colors

(b) *bright line, or emission, spectrum:* a pattern of bright colored lines (different wavelengths)

(c) *dark line, or absorption, spectrum:* a pattern of dark lines across a continuous spectrum

3. *Atoms* are responsible for each type of spectrum. An atom is the smallest particle of a chemical element that can exist.

More than one hundred chemical elements have been identified, each with its own particular kind of atom. Each element's atoms have a nucleus, bearing a unique positive charge, circled by an equal number of negatively charged electrons. Atoms are normally electrically neutral. The electrons are confined to a set of allowed orbits of definite radius. Each element has its own unique set of allowed electron orbits.

An electron in a particular orbit has a definite energy. An undisturbed atom, in the *ground state,* has the least possible energy. If energy is supplied to the atom, an electron may jump to a higher energy orbit. When the electron falls back down, the atom radiates energy in the form of a pellet of light called a *photon.*

If an atom absorbs enough energy, one or more of its electrons can be removed completely. The atom, which will then have a positive electric charge, is then called an *ion.*

Bright-colored *emission lines* are produced when electrons jump from higher orbits back down to lower orbits. Since each kind of neutral or ionized atom has its own unique set of orbits, each chemical element has its own unique set of bright-colored emission lines. For example, figure 3.4 shows the unique set of bright-colored emission lines for hydrogen.

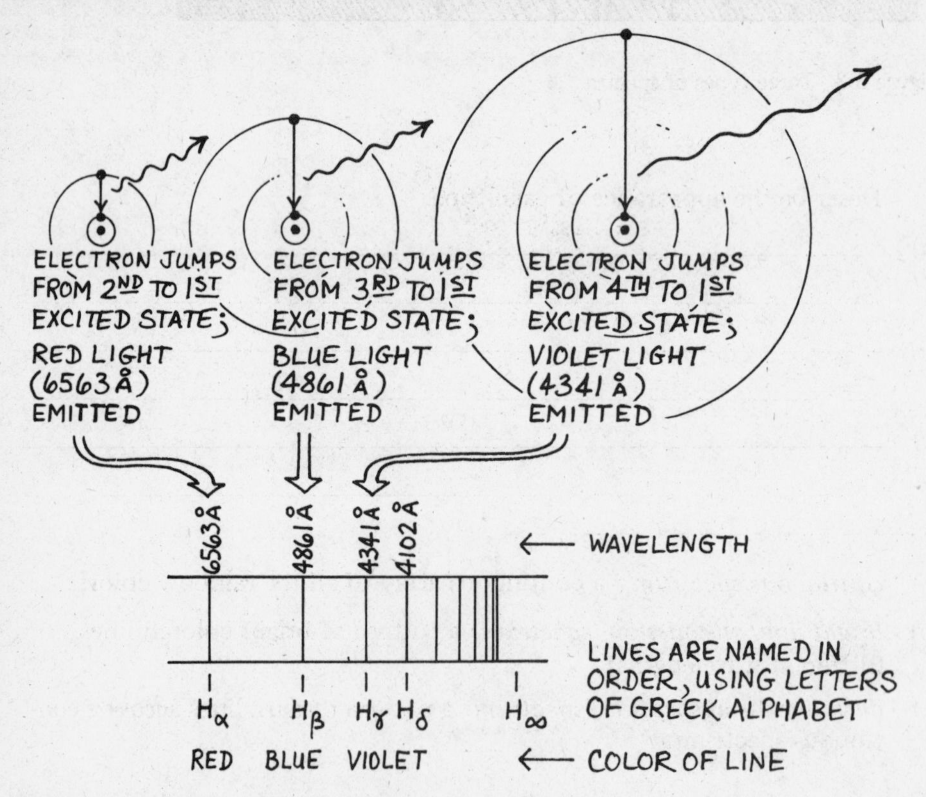

Figure 3.4. Origin of bright emission lines of hydrogen.

Corresponding *dark absorption lines* are produced when the atom absorbs light and the electrons jump out to higher orbits. For example, two dark absorption lines for hydrogen are indicated in figure 3.5.

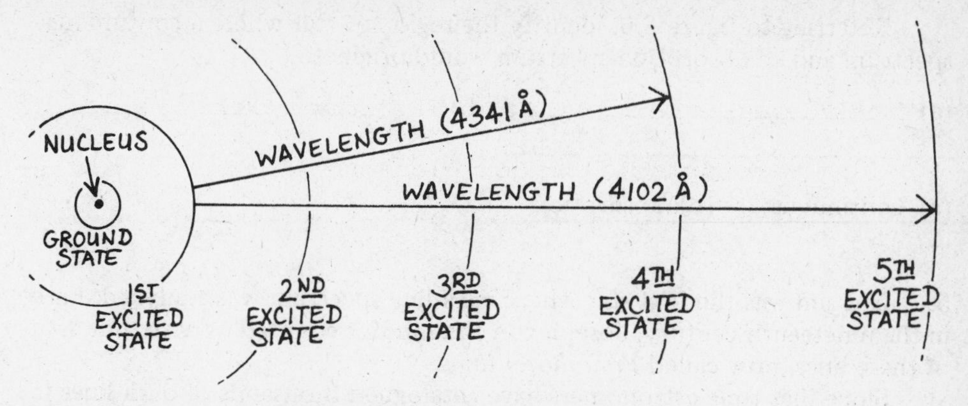

Figure 3.5. Origin of dark absorption lines of hydrogen.

Why do atoms emit light of different colors (specific wavelengths)?

Each color (wavelength) corresponds to an electron jumping down from a particular higher orbit to a particular lower one.

4. *Stellar spectra*, or the spectrums of stars, are predominantly patterns of dark lines crossing a continuous band of colors. (See frame 6.)

Stars are blazing balls of gas where many kinds of atoms emit light of all colors. Light from the star's opaque surface, called its *photosphere*, is blurred into a *continuous spectrum* of colors.

The light from the star's photosphere must pass through the star's *atmosphere* before shining into space. Atoms in the cooler atmosphere absorb some of the colors, producing dark absorption lines, which identify the chemical elements that make up the star's atmosphere.

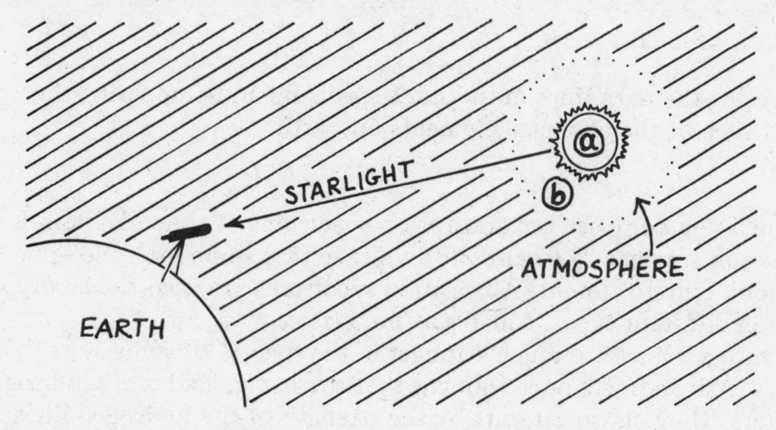

Figure 3.6. Spectra from a star.

Referring to figure 3.6, identify the region of star where a continuous spectrum and an absorption spectrum would originate.

(a) ______________________; (b) ______________________

(a) continuous; (b) absorption

5. Our sun was the first star whose dark line spectrum was analyzed. Early in the nineteenth century, Joseph von Fraunhofer counted over six hundred of these lines, now called *Fraunhofer lines.*

Since that time astronomers have catalogued thousands of dark lines in the sun's spectrum. By comparing them with the spectral lines produced by different chemical elements on Earth, they have found over sixty different chemical elements in the sun.

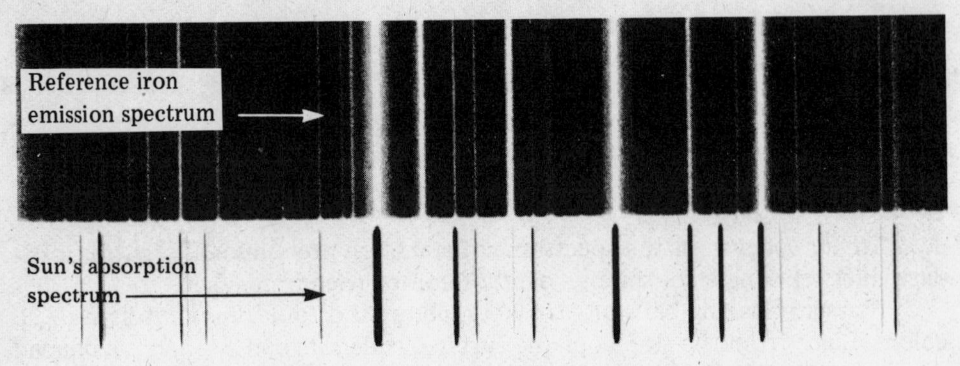

Figure 3.7. Sun's absorption spectrum compared to iron emission spectrum.

How can the chemical composition of stars be determined? (Assume that stars and their atmospheres are made of the same ingredients.) ______________________

____

---

by analyzing the dark lines in the star's spectrum and comparing them with those of each of the chemical elements on Earth

6. When you compare the absorption spectrums of stars like Polaris or Vega with the sun's spectrum (figure 3.8), you see that some look the same while others look quite different. Absorption spectrums are used to classify stars into seven different types, called *spectral classes.*

Hydrogen lines are much stronger in the spectra of some stars than in the sun's. Astronomers once thought that these stars had more hydrogen than other stars. They classified stars by the strength of the hydrogen lines in their spectra from the strongest (called Class A) to the weakest (called Class Q), in alphabetical order.

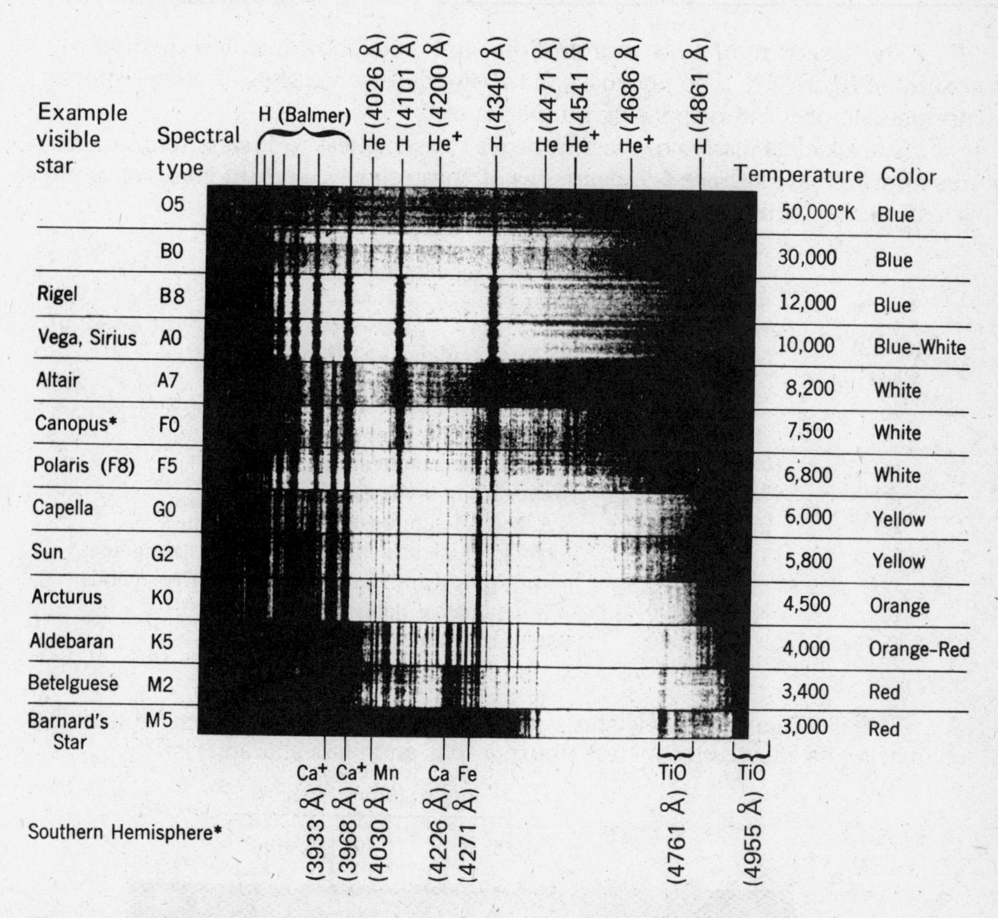

Figure 3.8. Seven classes of stellar spectra. The wavelengths of spectral lines are given in angstrom units (Å). (He: neutral helium; H: hydrogen; Ca: calcium; Fe: iron; TiO: titanium oxide; He^+: helium ion; Ca^+: calcium ion.)

Today we know that all visible stars are roughly uniform in composition. All are made mostly of hydrogen and helium. The differences in their dark line patterns are due primarily to their enormously different *surface temperatures.*

Stars are now classified in spectral classes corresponding to surface temperature, but the traditional identifying letters remain. When stars are arranged in order of decreasing temperature, from hottest to coolest, the spectral types are O B A F G K M. Astronomy students remember this strange order by saying: "*O*h *B*e *A* *F*ine *G*irl (*G*uy) *K*iss *M*e."

What property determines the spectral class of a star? ________________

____

---

surface temperature

7. The spectrum of a hot star and that of a cool star look very different. Examine figure 3.8. The photographs show the seven classes of stellar spectra, arranged in order of decreasing temperature.

The spectral classes of stars in order from highest to lowest temperatures, the approximate surface temperatures of these classes, and the main class characteristics are summarized in Table 3.1.

Table 3.1. Spectral class characteristics

Spectral Class	Approximate Temperature (°K)	Main Class Characteristics
O	>30,000	Hot stars with lines of ionized helium
B	10,000-30,000	Lines of neutral helium
A	7,500-10,000	Very strong hydrogen lines
F	6,000-7,500	Ionized calcium lines; many metal lines
G	4,500-6,000	Strong ionized calcium lines; many strong lines of ionized and neutral iron and other metals
K	3,500-4,500	Strong lines of neutral metals
M	2,000-3,500	Bands of titanium oxide molecules

You can identify a new star's spectral class and probable temperature by comparing its spectrum to these photographs and class characteristics.

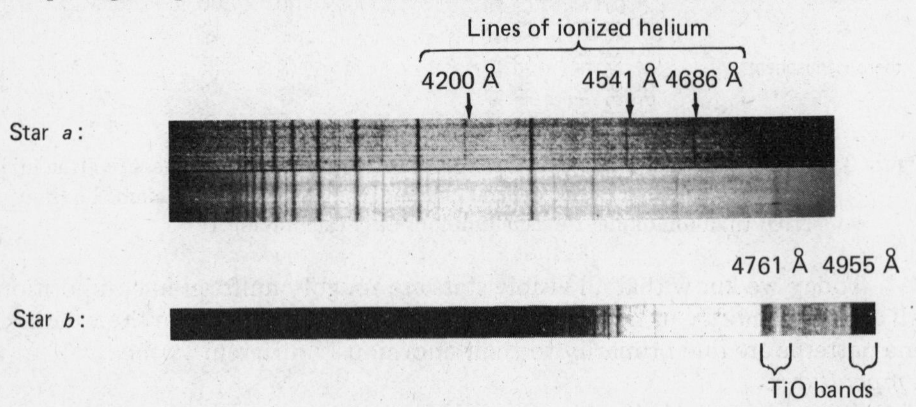

Figure 3.9. Star spectra.

List the spectral class and probable temperature of each of the stars whose spectrum is shown in figure 3.9. (a) ______________________;
(b) ______________________

(a) O type (> 30,000° K); (b) M type (2,000°–3,500° K)

8. Atomic theory explains why hot blue (O-type) stars and cool red (M-type) stars produce spectra that look so different even though all stars are made of practically the same ingredients.

Every chemical element has a characteristic temperature and density at which it is most effective in producing visible absorption lines.

At extremely high temperatures (as in O stars), gas atoms are *ionized*, or broken up. Only the tightest bound atoms such as singly ionized helium survive, and the lines of ionized atoms dominate the spectrum. When the temperature is lower (as in G stars, such as our sun), metal atoms such as iron and nickel remain neutral without being disrupted. At low temperatures (as in M stars), even molecules such as titanium oxide can exist.

Does the absence of the characteristic absorption lines of a particular element like hydrogen in a star's spectrum necessarily mean that the star does not contain that element? Explain. ______________________
____

No. The star's temperature determines which kinds of atoms can produce visible absorption lines.

9. Stars have a *space velocity* (motion through space with respect to the sun) of many miles a second. Space velocity has two components: *Radial velocity*, or speed toward or away from the sun, and *proper motion*, or the amount of angular change in a star's position per year.

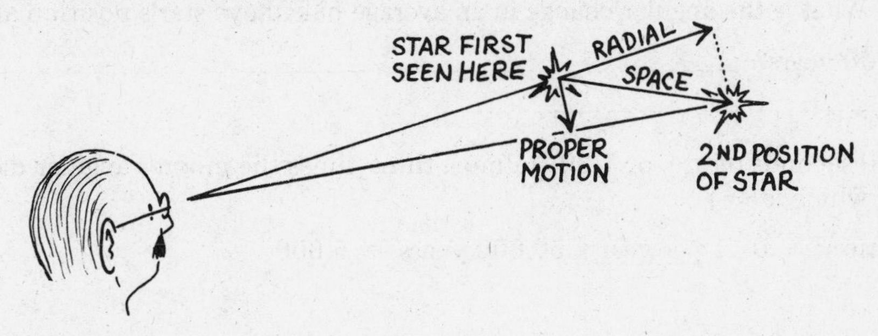

Figure 3.10. Space velocity.

A star's radial velocity is determined from analysis of its spectrum. The *Doppler shift* is an effect, discovered by Christian Doppler, that applies to all wave motion. When a source of waves and an observer are moving toward or away from each other, the wavelengths appear changed, or shifted.

A star's spectral lines (wavelengths) for any given element, often iron, are compared with a reference spectrum. (See figure 3.7.) If the star is moving toward us, these wavelengths look shorter (blue shift). If the star is moving away from us, its wavelengths look longer (*redshift*).

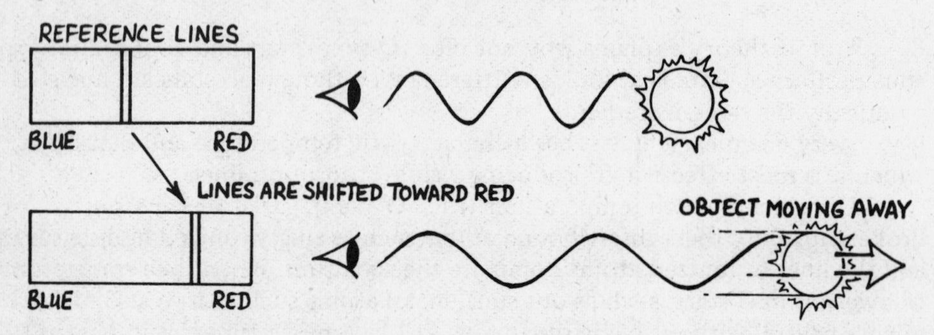

Figure 3.11. Redshift of spectral lines.

Proper motion is measured over an interval of 20 to 30 years. The average proper motion for all naked-eye stars is less than 0.1 second of arc ($0''.1$) per year. At that rate you won't notice any change in the appearance of your favorite constellation during your whole lifetime. But if you could return to observe the sky 50,000 years from now, it would look very different.

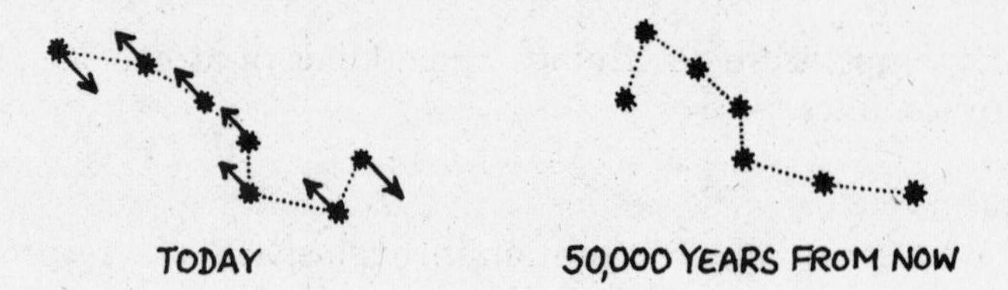

Figure 3.12. Proper motion of the Big Dipper.

What is the angular change in an average naked-eye star's position after 50,000 years? ______________________________

5,000 seconds of arc, or 1.39° (almost three times the moon's angular diameter, which is $\frac{1}{2}°$)

Solution: $0''.1$ per year × 50,000 years = $5{,}000''$

10. Other information about stars is obtained from careful measurements of *spectral line shape*.

Gas density is indicated by *collisional broadening.* A broadened spectral line is produced when atoms collide more frequently in higher density stars.

Axial rotation, the rotation of a star around its axis, is indicated by *rotational broadening.* If observable, a broadened line can yield a lower limit to the star's rate of rotation on its axis.

Average magnetic field strength is indicated by *magnetic broadening.* A splitting or broadening of spectral lines occurs in the presence of a magnetic field.

These different kinds of broadening are not distinguishable to the naked eye but are determined by careful analysis of the shape of the line using sensitive photometers.

List three properties of a star connected to its spectral line shape.

____

\- - - - - - - - - - - - - - - - - - -

(1) density; (2) axial rotation; (3) magnetic field strength

11. Write a brief paragraph summarizing your understanding of how astronomers deduce different properties of a star from its spectrum.

____

____

____

____

\- - - - - - - - - - - - - - - - - - -

Your summary should include the following concepts: (1) chemical composition, from the presence of the characteristic lines of certain elements; (2) temperature, from spectral type; (3) star's speed toward or away from us, from Doppler shift of the lines; (4) density, axial rotation, and surface magnetic fields, from line shape

12. ★ Astronomers distinguish between a star's *apparent brightness*—the way the star *appears* in the sky—and its *absolute luminosity*—the *actual* amount of light it shines into space each second.

The star we know best is our sun. The absolute luminosity of other stars is often stated in terms of the sun's luminosity, which is 3.83×10^{33} ergs per second. The sun's luminosity is equivalent to 3,830 billion trillion 100-watt light bulbs shining all together.

The most luminous stars are almost 100,000 times as luminous as the sun. The dimmest stars known are only one millionth as luminous as the sun.

Rigel, in Orion, is about 60,000 times as luminous as the sun. Can you explain why the sun looks much brighter than Rigel?

____

- - - - - - - - - - - - - - - - - -

Rigel is much farther away from Earth (815 light-years) than the sun is (about 93 million miles). A star's apparent brightness depends on both its luminosity and its distance away from us.

13. You cannot tell by looking at stars in the sky which ones have the greatest luminosity. Light spreads out uniformly in all directions from a source so that the amount of starlight shining on a unit area falls off as the square of the distance away from the star.

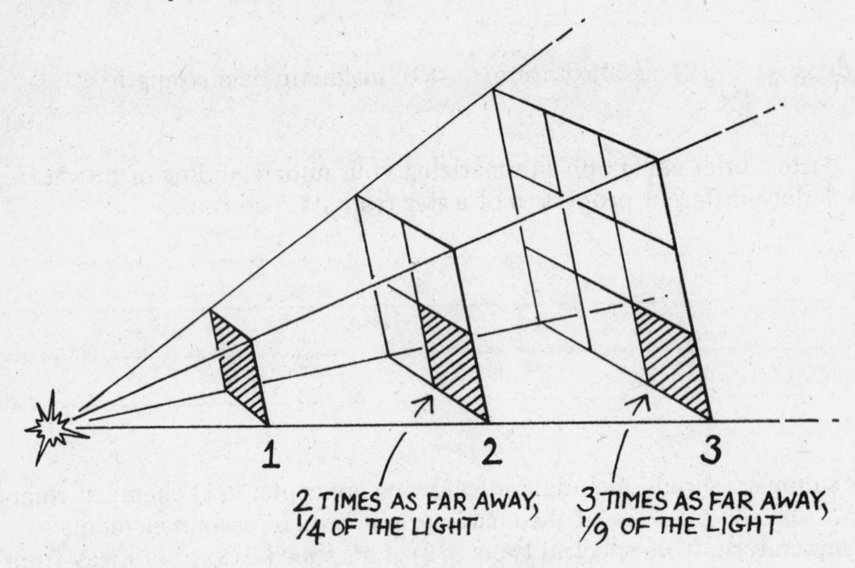

Figure 3.13. Decrease in apparent brightness with distance.

Thus if two stars have exactly the same luminosity, but one is twice as far away from you as the other, the distant one will look only $\frac{1}{2^2} = \frac{1}{4}$ as bright as the closer one, because you get one-fourth the light in your eyes.

Our sun is exceptionally bright, because it is so close to us. If it were located 100,000 times deeper in space, how many times fainter would it look?

- - - - - - - - - - - - - - - - - -

10 billion times fainter, or about like the brilliant blue-white star, Sirius.

Solution: $\frac{1}{(100{,}000)^2} = \frac{1}{10{,}000{,}000{,}000}$ as bright (or 10 billion times fainter).

---

14. *Apparent magnitude* is a measure of how bright a star appears. (See frame 1.7.) The modern *magnitude scale* defines a first-magnitude star to be exactly one hundred times brighter than a sixth-magnitude star.

This ratio agrees with the way our eyes respond to increases in brightness of stars. What we see as a linear increase in brightness (1 magnitude difference) is precisely measured as a geometrical increase in brightness (2.51 times brighter).

Magnitude differences between stars measure the relative brightness of the stars. Table 3.2 lists approximate brightness ratios corresponding to sample magnitude differences.

Table 3.2. Magnitude differences and brightness ratios

Difference in Magnitude	Brightness Ratio
0.0	1:1
1.0	2.5:1
2.0	6.3:1
3.0	16:1
4.0	40:1
5.0	100:1
6.0	251:1
10.0	10,000:1
15.0	1,000,000:1
20.0	100,000,000:1
25.0	10,000,000,000:1

Remember that the most negative magnitude numbers identify the brightest objects, while the largest positive magnitude numbers identify the faintest objects. (See frame 1.7.)

Consult Table 3.3.

Table 3.3. Sample magnitude data

Subject	Apparent Magnitude
Sun	-26.7
Venus (at brightest)	- 4
Sirius	- 1.5
Antares	1
Naked-eye limit	6.5
Binocular limit	10
6-inch telescope limit	13
200-inch (visual) limit	20
200-inch photographic limit	23.5

How much brighter does the sun appear than Sirius? Explain. ______________

____

____

----

10 billion times brighter

Solution: Magnitude difference is (-26.7) - (-1.4) ≅ 25, corresponding to brightness ratio of 10,000,000,000:1.

15. *Absolute magnitude* is a measure of luminosity, or how much light a star is actually radiating into space. If you could line up all stars at the same distance from Earth, you could see how they differ in intrinsic, or "true," brightness.

Astronomers define a star's absolute magnitude as the apparent magnitude it would have if it were located at a standard distance of 10 parsecs from us.

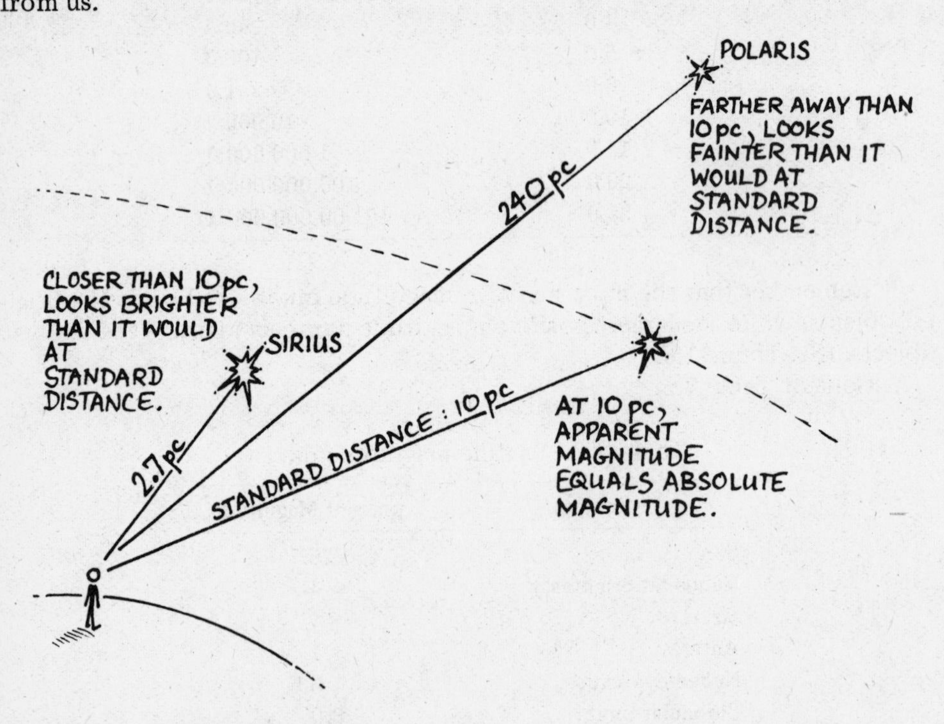

Figure 3.14. Absolute magnitude.

If a star is farther than 10 parsecs away from us, its apparent magnitude is numerically bigger than its absolute magnitude. (Large positive magnitude numbers indicate faint objects.) For example, *Polaris* is 240 pc away. Its apparent magnitude is +2.3, whereas its absolute magnitude is -4.6.

On the other hand, if a star is closer than 10 parsecs, its apparent magnitude is numerically smaller than its absolute magnitude. Thus, *Sirius* is 2.7 pc away; its apparent magnitude is −1.5, whereas its absolute magnitude is only +1.4.

Consider the two bright stars Deneb and Vega. Refer back to Table 1.1 to fill in the chart below.

Star	Constellation	Apparent Magnitude	Absolute Magnitude
Deneb	Cygnus	(a)	(b)
Vega	Lyra	(c)	(d)

Then tell: (a) which looks brightest? __________ (b) Which is really most luminous? __________ (c) What factor makes your answers to (a) and (b) different? ______________________________

Chart: (a) 1.3 (b) −7.3 (c) 0.0 (d) 0.5

(a) Vega (numerically smallest apparent magnitude)
(b) Deneb (numerically most negative absolute magnitude)
(c) the star's distance away from us

16. The difference between the apparent magnitude (m) and absolute magnitude (M) is called the *distance modulus.* In formula form,

$$m - M = 5\log\left(\frac{\text{distance in parsecs}}{10}\right).$$

A star's apparent magnitude can be measured directly.

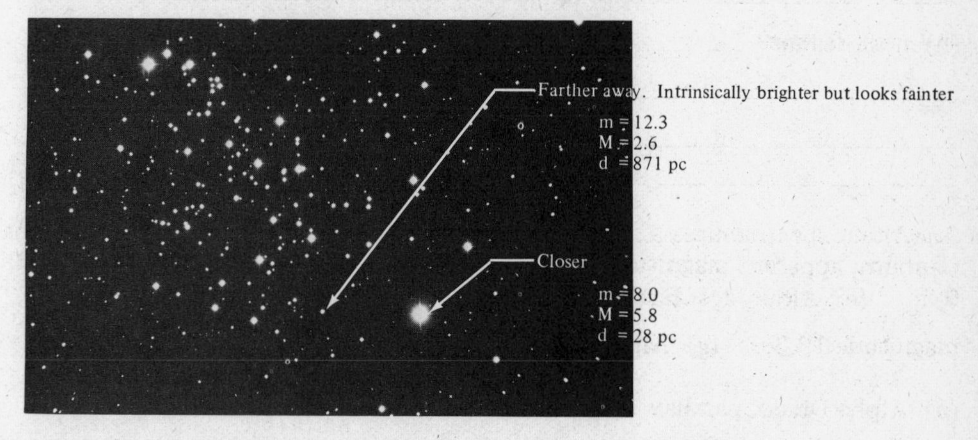

Figure 3.15. Star cluster, showing difference in apparent and absolute magnitude. Farther away (intrinsically brighter but looks fainter): $m = 12.3$; $M = 2.6$; $d = 871$ pc. Closer: $m = 8.0$; $M = 5.8$; $d = 28$ pc.

For a very distant object whose parallax cannot be measured but whose absolute magnitude is known, as from consideration of its spectrum, the formula can be used to calculate distance.

Give the distance modulus, as indicated in figure 3.15, of the stars which are (a) closer ________; (b) farther away ________.

- - - - - - - - - - - - - - - - - -

(a) 2.2; (b) 9.7

17. Make sure you understand the ideas presented so far by answering the following questions about four of the sun's neighbor stars described in Table 3.4.

Table 3.4. Four nearby stars

Star	Apparent Magnitude	Absolute Magnitude	Spectral Class	Parallax (in ″)
Alpha Centauri	0.0	4.4	G	0″.745
Alpha Draco	4.7	5.9	K	0.176
Barnard's Star	9.5	13.3	M	0.552
Altair	0.8	2.2	A	0.197

Which star is (a) hottest? ________ (b) coolest? ________ (c) brightest looking ________ (d) faintest appearing? ________ (e) intrinsically (actually) most luminous? ________ (f) intrinsically least luminous? ________ (g) closest? ________ (h) most distant? ________ Explain your answers.

- - - - - - - - - - - - - - - - - -

(a) Altair, spectral class A; (b) Barnard's Star, spectral class M; (c) Alpha Centauri, apparent magnitude 0.0; (d) Barnard's Star, apparent magnitude 9.5; (e) Altair, absolute magnitude 2.2; (f) Barnard's Star, absolute magnitude 13.3; (g) Alpha Centauri, distance $= \frac{1}{p} = \frac{1}{0''.745} = 1.3$ pc; (h) Alpha Draco, parallax = 0″.176, or distance $= \frac{1}{0''.176} = 5.7$ pc

18. The *Hertzsprung-Russell* (H-R) *diagram* is an important graph that shows luminosity versus temperature for many stars. Astronomers use the H–R diagram to check their theories of stellar evolution (discussed in Chapter Four) and the internal structure of stars.

Every dot on an H-R diagram represents a star whose temperature (spectral class) is read on the horizontal axis and whose luminosity (absolute magnitude) is read on the vertical axis.

Significantly, when a few thousand stars are chosen randomly and plotted on an H-R diagram, they fall along definite tracks. This pattern suggests that a meaningful connection exists between a star's luminosity and its temperature. Otherwise, the dots would be scattered randomly all over the graph.

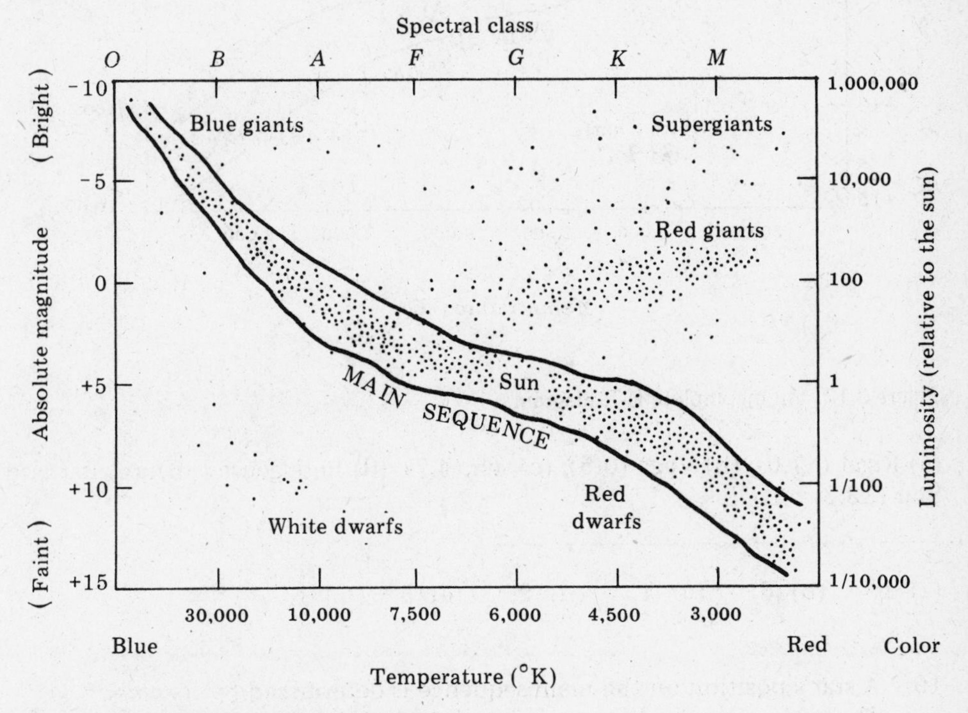

Figure 3.16. Hertzsprung-Russell diagram.

About 90 percent of the stars lie along a band called the *main sequence* that runs from the upper left (hot, very luminous *blue giants*), across the diagram to the lower right (cool, faint *red dwarfs*). Red dwarfs are the most common type of nearby star.

Most of the other 10 percent of stars fall in the upper right region (cool, bright *giants* and *supergiants*) or in the lower left corner (hot, low-luminosity *white dwarfs*).

Identify the location of the following stars indicated on the H–R diagram in figure 3.17. (The stars absolute magnitude is given in parentheses.) Refer to figure 3.8 for temperature and spectral class.

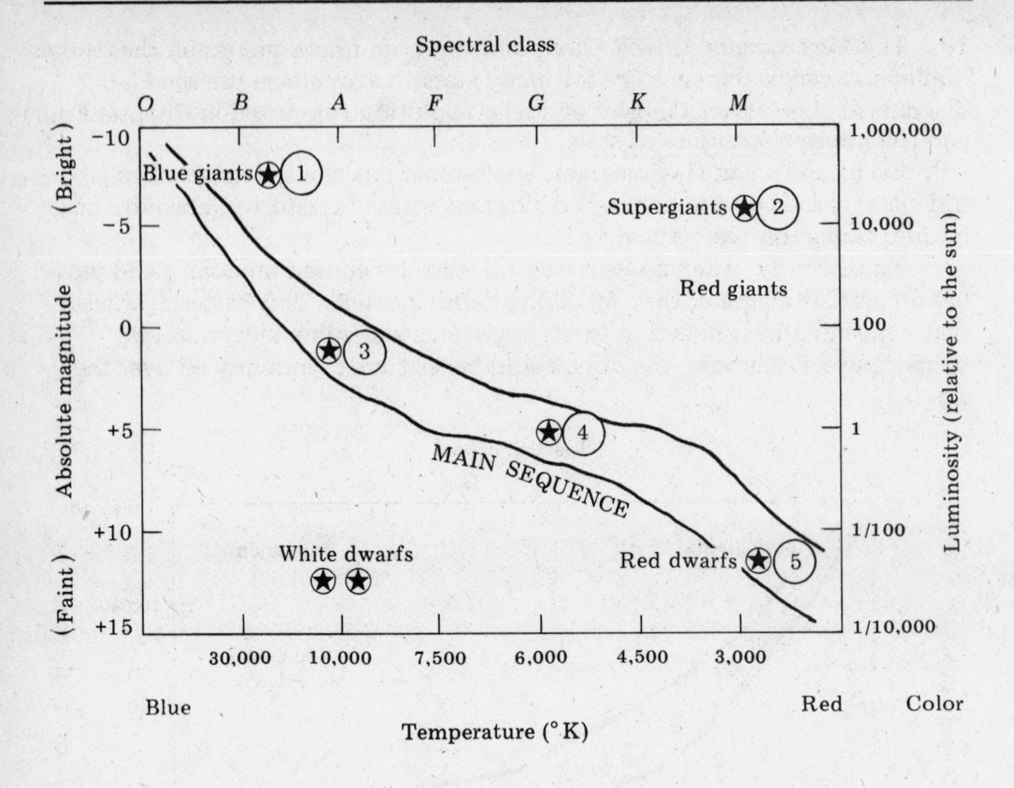

Figure 3.17. An incomplete H-R diagram.

(a) Rigel (-7.0); (b) Vega (0.5); (c) sun (4.7); (d) Betelgeuse (-6); (e) Barnard's Star (13.3)

(a) 1; (b) 3; (c) 4; (d) 2; (e) 5

19. A star's position on the main sequence is determined by its *mass.*

The main sequence is a sequence of stars of decreasing mass from the most massive, most luminous stars at the upper end to the least massive, least luminous stars at the lower end.

An empirical *mass luminosity relationship*, found from binary stars, says that the more massive stars are also the most luminous. The luminosity of a star is approximately proportional to its mass raised to the 3.5 power.

The mass of the sun is 2×10^{33} grams, over 300,000 times the mass of Earth. Stellar masses do not vary enormously along the main sequence as stellar luminosities do. The faintest red dwarfs have a mass about one-tenth of the sun's, while the largest mass of a stable star is about sixty times that of the sun.

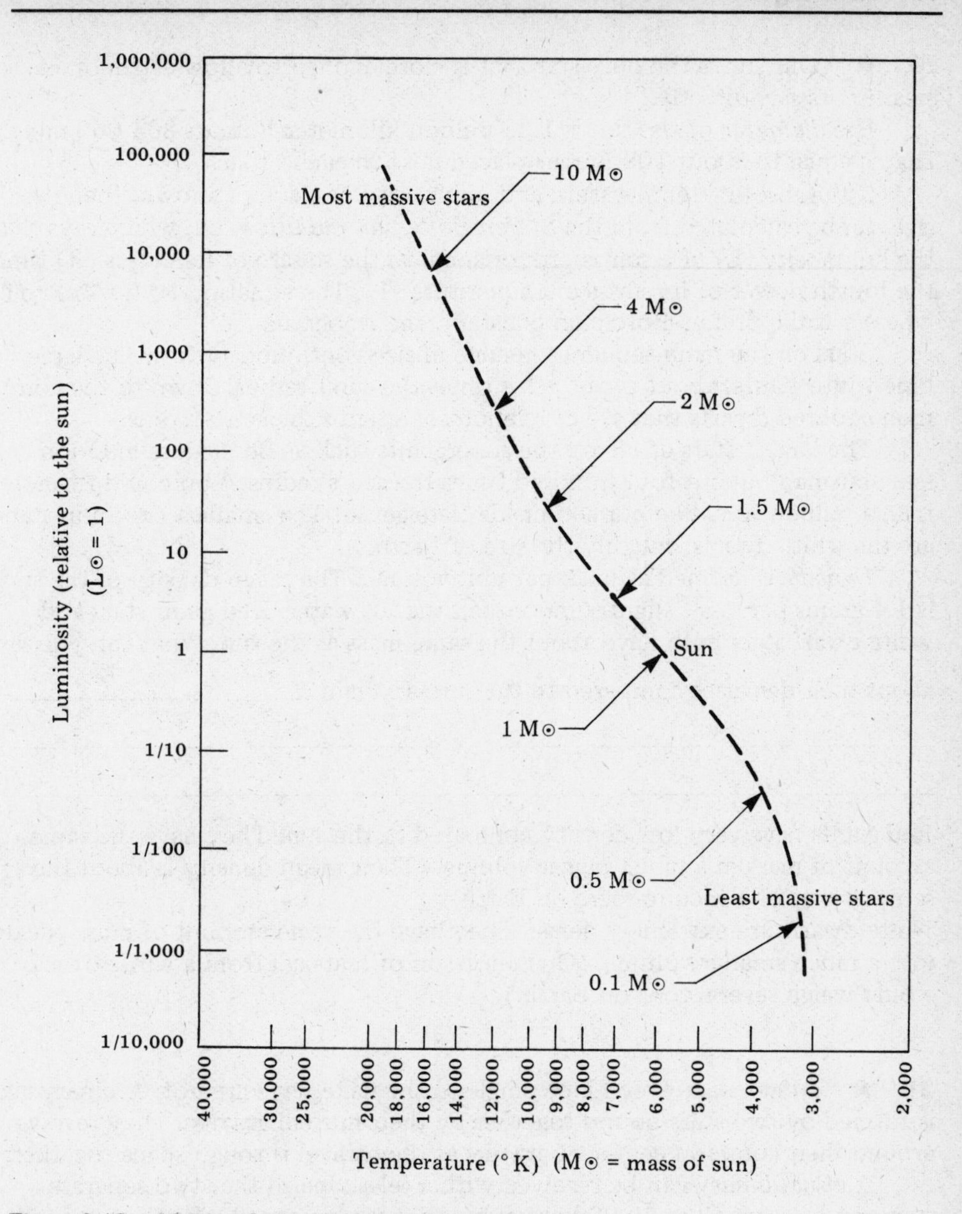

Figure 3.18. Masses of some main sequence stars.

What basic property of a star determines its position on the main sequence of the H–R diagram; that is, what determines its luminosity and temperature? ____________________

- - - - - - - - - - - - - - - - - -

mass

20. ★ Our sun is the only star that is close enough to allow astronomers to measure its *size* directly.

The *diameter* of the sun is 1.39 million kilometers (about 864,000 miles). That is equal to about 109 Earths placed next to each other.

If the absolute temperature and luminosity of a star is known, then its size can be calculated from the *Stefan-Boltzman radiation law,* which says that the luminosity (L) of a star is proportional to the square of its radius (R) times the fourth power of its surface temperature (T). The equation is: $L = 4\pi R^2 \sigma T^4$, where σ is the Stefan–Boltzman constant (see Appendix 2).

Stars on the main sequence change in size continuously from the large blue-white giants, about twenty-five times the sun's radius, down to the common cool red dwarfs that are only about one tenth the sun's radius.

The largest stars of all are the supergiants such as Betelgeuse in Orion (see star map), about four hundred times the sun's radius. You could fit more than a million stars like our sun inside Betelgeuse! The smallest common stars are the white dwarfs, roughly the size of Earth.

Density is defined as mass per unit volume. The mean density of the sun is 1.4 grams per cm^3, slightly more than that of water. Red giant stars and white dwarf stars both have about the same mass as the sun. What can you say about their densities compared to the sun? Explain ______________________

____

____

Red giants have very low density compared to the sun. They have the same amount of mass in a much bigger volume. (Their mean density is about the same as that of a vacuum here on Earth.)
White dwarfs are extremely dense. They have the same amount of mass packed into a much smaller volume. (One teaspoon of material from a white dwarf would weigh several tons on Earth.)

21. ★ Many stars which look single to the naked eye are not. A binary star is formed by two stars bound together by their mutual gravity. They revolve around their common center of gravity as they travel through space together.

A *visual binary* can be resolved with a telescope so that two separate stars can be seen. Over 70,000 visual binaries are known. Alpha Centauri, our closest visible neighbor star, is actually a double star. Nearby is faint Proxima Centauri, the third member of this multiple star system.

Two interesting visual binaries can easily be seen in a small telescope. Mizar in the Big Dipper was the first binary star discovered (in 1650). Beautiful Alberio in Cygnus is a colorful yellow and blue star (see star maps). Many others are listed in observer's guides. (See Useful References.)

Many visible stars may have companions that are too faint to be seen. The presence of an unseen companion is inferred from variable proper motion of the visible companion. An *astrometric binary* is a visible star plus an unseen companion. Brilliant Sirius (Sirius A) in Canis Minor was an astrometric binary from 1844, when its nature was detected, until 1862. Then its faint companion (Sirius B) was observed.

A *spectroscopic binary* cannot be resolved in a telescope. Its binary nature is revealed by its spectrum. A Doppler shift is apparent in the spectral lines as the companions approach and recede from Earth. About eight hundred spectroscopic binaries have been analyzed. The brighter member of Mizar (Mizar A) is a spectroscopic binary.

An *eclipsing binary* is situated so that one star passes in front of its companion, cutting off light from our view at regular intervals. An eclipsing binary regularly changes in brightness. You can see with your naked eye the famous eclipsing binary Algol, the demon, in Perseus. Algol "winks" from brightest magnitude 2.2 to least bright magnitude 3.5 in about 2 days and 21 hours.

An *optical double* is a pair of stars that appear to be close to one another in the sky when viewed from Earth. Actually one is much more distant than the other, and they have no physical relationship to one another.

Test your eyesight by finding both Mizar and Alcor (nicknamed "the testers"), the optical double in the handle of the Big Dipper.

How does an optical double differ from a visual binary? ____________

____

____

— — — — — — — — — — — — — — — — — — —

The stars in an optical double are far apart and have no actual physical relationship to one another. The stars in a visual binary are bound together in space by their mutual gravity.

SELF-TEST

This self-test is designed to show you whether or not you have mastered the material in Chapter Three. Answer each question to the best of your ability. Correct answers and review instructions are given at the end of the test.

1. Refer to Table 3.4. From the measured parallax, find the distance to Barnard's Star in (a) parsecs ____________; (b) light-years ____________.

2. Explain why the bright (dark) spectral lines of light emitted from (absorbed by) the atoms of an element are unique to that element.

   ____

   ____

3. Explain how a star's spectrum is formed. ______________________________

   ____

4. List the following types of spectral lines in order as they appear in stars of decreasing temperature.

   ___ (1) very strong hydrogen lines
   ___ (2) ionized helium
   ___ (3) bands of titanium oxide molecules
   ___ (4) neutral helium
   ___ (5) neutral metals
   ___ (6) ionized metals

   ________ ________ ________ ________ ________ ________

5. Match the following properties of a star that can be deduced from its spectrum with the appropriate method listed on the right.

   ___ (a) chemical composition
   ___ (b) temperature
   ___ (c) radial velocity
   ___ (d) gas density, axial rotation, magnetic field

   1. Doppler shift
   2. spectral type (class)
   3. line shape
   4. characteristic lines

6. The proper motion of Sirius is 1.32″ per year. Find how much Sirius will change its position on the celestial sphere in the next 1,000 years.

   ____

7. Define space velocity. ______________________________________

   ____

8. Refer to Table 1.1. By using their apparent magnitudes, absolute magnitudes, and spectral classes, match one of the four stars to each description.

___ (a) hottest	1. Betelgeuse
___ (b) coolest	2. Canopus
___ (c) most luminous	3. Rigel
___ (d) least luminous	4. Sirius
___ (e) brightest	
___ (f) faintest	
___ (g) closest	
___ (h) most distant	

9. Label the following on the H–R diagram below:

1. surface temperature of star (°K)	6. red giants
2. absolute luminosity (sun = 1)	7. white dwarfs
3. spectral class	8. supergiants
4. absolute magnitude	9. blue giants
5. main sequence	10. red dwarfs

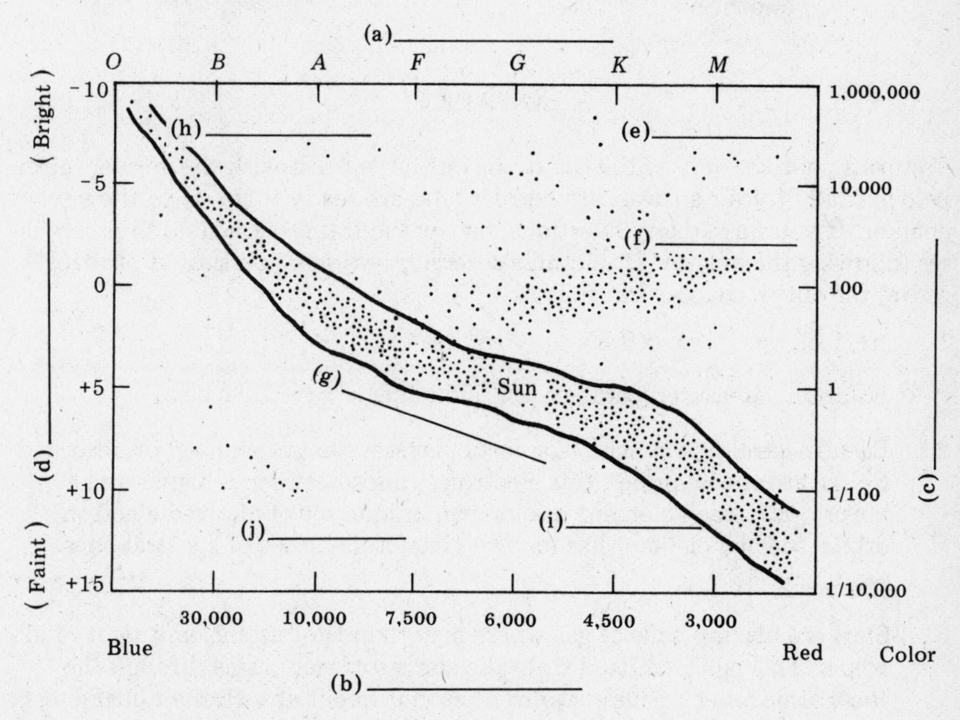

10. What is the most basic property of a star that determines its location on the main sequence (temperature and luminosity)? ______________

____

11. Use the H–R diagram to explain why, compared to our sun, red giants must be very large and white dwarfs must be very small. ______________

____

____

12. Match:

___ (a) can be resolved with a telescope

___ (b) unseen companion inferred from variable proper motion of visible companion

___ (c) binary nature revealed by its spectrum

___ (d) changes in brightness regularly as one star blocks its companion from our view

___ (e) member stars have no actual physical relationship to one another

1. astrometric binary
2. eclipsing binary
3. optical double
4. spectroscopic binary
5. visual binary

ANSWERS

Compare your answers to the questions on the self-test with the answers given below. If all of your answers are correct, you are ready to go on to the next chapter. If you missed any questions, review the frames indicated in parentheses following the answer. If you miss several questions, you should probably reread the entire chapter carefully.

1. (a) 1.81 pc; (b) 5.9 LY (frame 1)

   Solution: Measured parallax is $0''.552$ and $\frac{1}{0''.552} = 1.81$ pc.

2. Each spectral line is light of a particular wavelength emitted or absorbed by the atom when one of its electrons jumps between a higher and a lower orbit. Each element has its own unique set of allowed electron orbits, so each element has its own characteristic set of spectral lines. (frames 2, 3)

3. Stars are blazing balls of gas where many kinds of atoms emit light of all colors. This light, emitted from the star's surface, passes through the star's atmosphere. There, atoms of each element absorb their characteristic wavelengths, so a pattern of dark lines crosses the continuous band of colors—the star's spectrum. (frames 3, 4)

4. 2; 4; 1; 6; 5; 3 (frame 7)

5. (a) 4 (b) 2 (c) 1 (d) 3 (frames 3, 6, 7, 9, 10)

6. 1320″, or roughly one-third of a degree

   Solution: Proper motion = 1.32″ per year. A degree is equal to 3600″.
   (1.32″ per year) x 1,000 years (frame 9)

7. velocity of a star with respect to the sun (frame 9)

8. (a) 3; (b) 1; (c) 3; (d) 4; (e) 4; (f) 1; (g) 4; (h) 3

   Solution:

	Spectral Class	Apparent Magnitude	Absolute Magnitude
Betelgeuse	M	+0.8	-6.0
Canopus	F	-0.7	-4.7
Rigel	B	+0.1	-7.0
Sirius	A	-1.5	+1.4

   (frames 6, 7, 12, 13, 14, 15, 16)

9. (a) 3; (b) 1; (c) 2; (d) 4; (e) 8; (f) 6; (g) 5; (h) 9; (i) 10; (j) 7

   (frame 18)

10. mass (frame 19)

11. Red giants are relatively cool but luminous; hence, they must have a large surface area radiating energy. White dwarfs are relatively hot but faint; hence, they must have a small surface area radiating energy into space. (frames 18, 19, 20)

12. (a) 5; (b) 1; (c) 4; (d) 2; (e) 3 (frame 21)

CHAPTER FOUR

Stellar Evolution

> To every thing there is a season, and a time to every purpose under the heaven.
> A time to be born, and a time to die.
>
> Ecclesiastes 3:1-2

1. No star shines forever. *Stellar evolution* refers to the changes that take place in stars as they age—the life cycle of stars. These changes cannot be observed directly, because they often take place over billions of years. Astronomers construct a theory of stellar evolution that is consistent with the laws of physics. Then they check their theory by observing real stars shining in the sky.

In checking theory against observations, astronomers make use of H-R diagrams. Theoretical predictions are made regarding a sequence of changes in luminosity and temperature for stars as they go from birth to death. These changes are plotted on an H-R diagram, forming theoretical *tracks of evolution.* Theoretical H-R diagrams are then compared with H-R diagrams for groups of real stars. (See also frame 5.9.)

The predictions of the modern theory of stellar evolution, described in this chapter, agree well with the data from observations of real stars.

What is stellar evolution? ____

____

- - - - - - - - - - - - - - - - - -

The changes that take place in stars as they age—the life cycle of stars.

2. ★ Stars form out of matter that exists in space. The interstellar (between stars) clouds of gas and dust must be the birthplaces of stars.

You can see for yourself a cloud in space where new stars are forming right now. The famous *Orion Nebula* (cloud), located about 1,500 light-years away in the constellation Orion, is a region of intense star formation.

Figure 4.1. Orion Nebula, in the constellation Orion.

Look for the Orion Nebula in the winter. It is marked on your star map in the sword of Orion the Hunter.

The Orion Nebula looks like a hazy patch to your naked eye. Through a telescope you will see it glow with a greenish color. Astronomers say that hot, newly formed stars in the region make the gases glow.

Are new stars still being born today? Where? ______________________

____

- - - - - - - - - - - - - - - - - - - -

Yes. In clouds of gas and dust, as in the Orion Nebula (figure 4.1).

3. A *protostar* is a huge, whirling, contracting, gaseous cloud. You can think of a protostar as a star that is being born. Protostars form by chance inside the turbulent clouds of gas (mostly hydrogen) and dust that exist in space.

A protostar is held together by the force of gravity. The force of gravity pulls matter in toward the center of the protostar, causing it to contract. Gravitational contraction causes the temperature and pressure of the gas inside to rise greatly.

Heat flows from the hot center to the cooler surface. The protostar radiates this energy into space. It glows with a deep red color.

When the temperature in the center reaches 10 million degrees Kelvin, *nuclear fusion* reactions start. These nuclear reactions release tremendous amounts of energy. Energy is generated in the center as fast as it is being radiated out into space. The very high internal temperatures and pressures are thus maintained.

The outward pressure of the very hot gases balances the inward pull of gravity. The protostar stops contracting. It shines its own light steadily into space. The protostar becomes a newborn star. Our sun was born in this way about 5 billion years ago.

List the three main steps in the birth of a star. ____________________

____

____

- - - - - - - - - - - - - - - - - -

1. gravitational contraction of a cloud of gas and dust
2. rise in interior temperature and pressure
3. nuclear fusion

4. The clouds where protostars form do not have identical masses or distributions of the chemical elements. The life cycle of a star—the time that it takes for a star to evolve—depends on its *mass* and *chemical composition.*

Stars that begin life with about the same mass and chemistry go through the same stages of evolution in about the same amount of time.

Stars of similar chemistry with very high mass evolve fastest, while those of very low mass take the longest time to evolve.

The theoretical evolutionary tracks on the H-R diagram in figure 4.2 show how a protostar's luminosity and temperature change as it contracts to become a star. Approximately how long does it take each of the following protostars to reach the zero age main sequence (to be born)? (a) stars like our sun ____________________ (b) stars with mass much greater than the sun's ____________________ (c) stars with mass much less than the sun's ____________________

- - - - - - - - - - - - - - - - - -

(a) about 30 million years; (b) about 100,000 years; (c) about 100 million years

5. You can think of a *main sequence star* as an adult star. In comparison to changes in protostars, evolution of main sequence stars is very slow. A star spends most of its lifetime shining steadily, with luminosity and temperature values found along the main sequence of H-R diagrams. A main sequence star gets its energy from nuclear fusion reactions in which hydrogen at the center of the star is converted into helium. Four hydrogen nuclei are fused together into one lighter helium nucleus. The disappearing mass is changed into energy and released. (The same process releases energy in hydrogen bombs.)

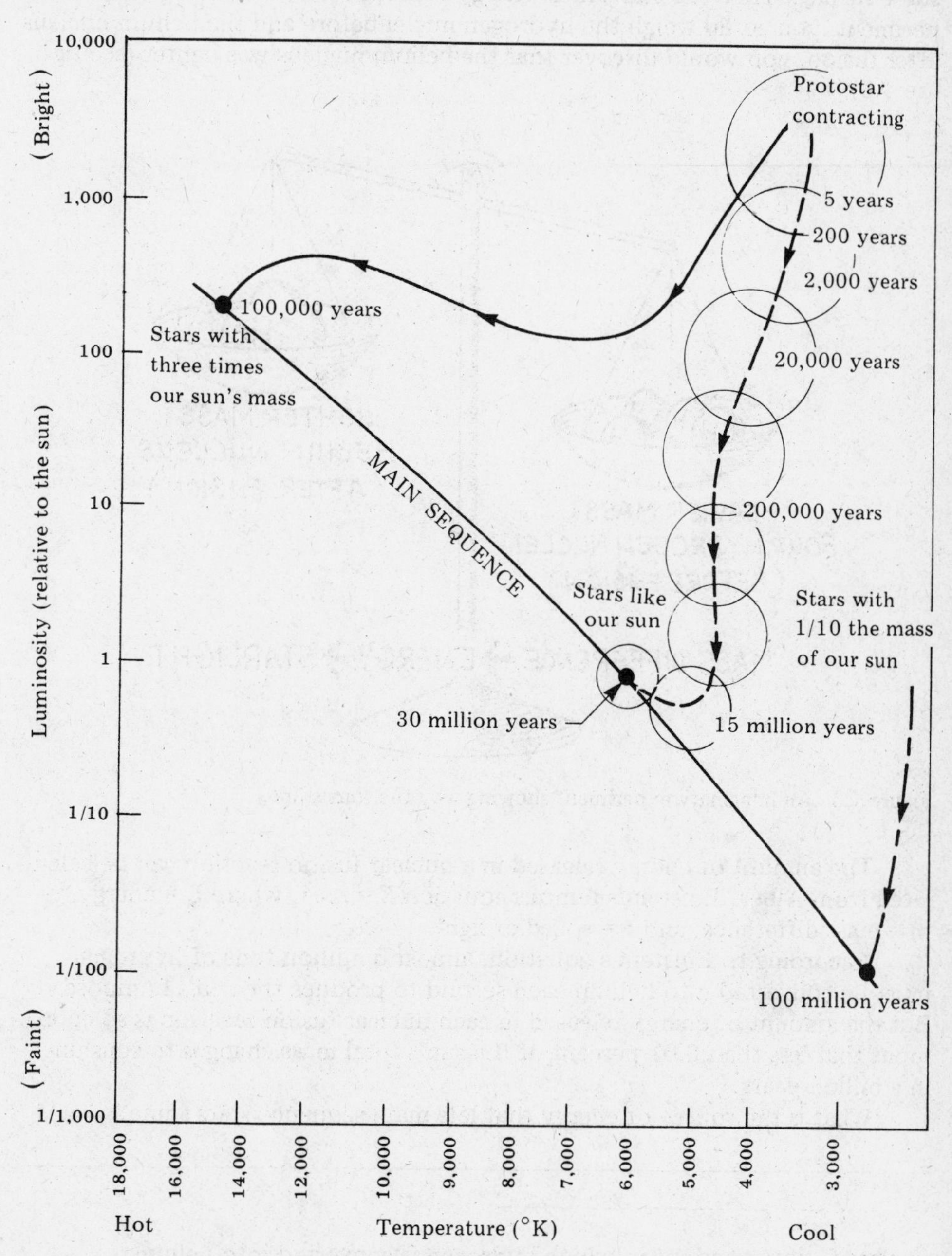

Figure 4.2. Luminosity and temperature change of contracting protostar.

The energy from the nuclear fusion reactions eventually reaches the star's surface. Then the star shines energy into space. If, in an imaginary experiment, you could weigh the hydrogen nuclei before and the helium nucleus after fusion, you would discover that the helium nucleus was lighter (see figure 4.3).

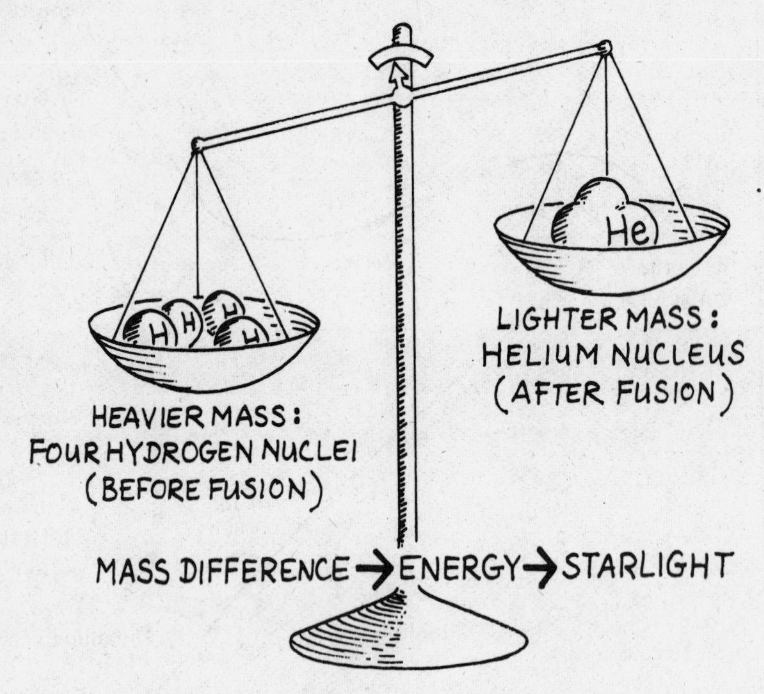

Figure 4.3. An imaginary experiment showing why the stars shine.

The amount of energy released in a nuclear fusion reaction can be calculated from Albert Einstein's famous equation $E = mc^2$, where E = energy, m = mass difference, and c = speed of light.

According to Einstein's equation, almost 5 million tons of hydrogen must be converted into helium each second to produce the sun's luminosity. But the amount of energy released in each nuclear fusion reaction is so enormous that less than 0.01 percent of the sun's total mass changes to sunshine in a billion years.

What is the source of energy that lets main sequence stars shine?

- - - - - - - - - - - - - - - - - - -

nuclear fusion reactions in which hydrogen is converted into helium

6. A star will shine steadily as a main sequence star until all the available hydrogen in its core has been converted into helium. Then the star will begin to die. (Review Chapter Three, frames 18 and 19.)

Our sun is an average medium-sized star. It has been shining as a stable main sequence star for about 5 billion years, and it should continue to shine steadily for another 5 billion years.

Very massive, hot, bright stars die fastest, because they use up their hydrogen most rapidly. The very massive blue giant stars spend only a few million years shining as main sequence stars.

The least massive, cool, dim stars live the longest because they consume their hydrogen fuel least rapidly. The small-mass red dwarfs are the oldest and most numerous main sequence stars. They have lifetimes billions of years long.

What type of stars are expected to live (a) longest? ________________
(b) shortest? ________________ (c) About how much longer is the sun expected to shine as it is now? ________________

(a) Those with small mass, such as red dwarfs (b) very massive stars, such as blue giants (c) about 5 billion years

7. ★ After the hydrogen fuel in the star's core is used up, the star no longer has an energy source in the core. The core, which then consists primarily of helium, begins to contract gravitationally. Hydrogen burning continues at the boundary between the helium core and the outside envelope of hydrogen.

Gravitational contraction of the helium core makes its temperature rise. The high temperature makes the boundary hydrogen burn faster, and the star's luminosity increases.

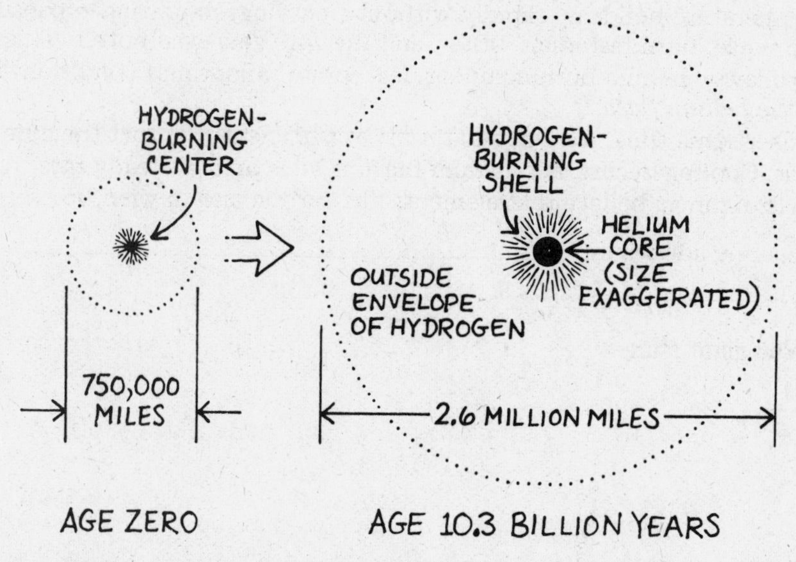

Figure 4.4. Sunlike star at start of life on main sequence and as it ages.

The tremendous energy released by the burning hydrogen and the gravitational contraction makes the star expand to gigantic proportions. The star's density is then very low everywhere except in the core.

As the star expands, its surface temperature drops. The star's surface color turns to red. The star has changed into a huge, bright, red, aging star—a *red giant.* It is cool but bright because of its gigantic surface area. It has the luminosity and temperature values of the red giant region of the H-R diagram.

You can see for yourself some red supergiant stars shining in the sky. Good examples are Betelgeuse and Antares, both over four hundred times the sun's diameter (refer to Table 2.1 and star maps).

Our sun, like all stars, is expected to change into a huge red giant when it dies. That red giant sun will shine so brightly that rocks will melt, oceans will evaporate, and life as we know it on Earth will end.

When does a star begin to change from a main sequence star into a red giant? ______________________________

— — — — — — — — — — — — — — — — — — —

When it has converted all of the available hydrogen fuel in its core into helium.

8. Gravitational contraction causes the temperature inside the red giant's helium core to rise to 100 million degrees Kelvin. At that temperature helium is converted to carbon in nuclear fusion reactions.

Inside the massive red giants, further fusion reactions can build up oxygen, sodium, magnesium, and other well-known elements heavier than helium. At still higher temperatures common metals such as iron are made.

The helium core does not expand much once the helium burning starts. The temperature builds up rapidly without a cooling, stabilizing expansion. Helium nuclei burn faster and faster, and the core gets even hotter. This onset of runaway helium burning under degenerate (abnormal) conditions is called the *helium flash.*

After some time, the temperature rises sufficiently so that the core expands. Cooling occurs, and helium burning goes on at a steady rate.

Astronomers believe that elements like carbon and oxygen, so essential for life, are made where? ______________________________

— — — — — — — — — — — — — — — — — — —

inside red giant stars

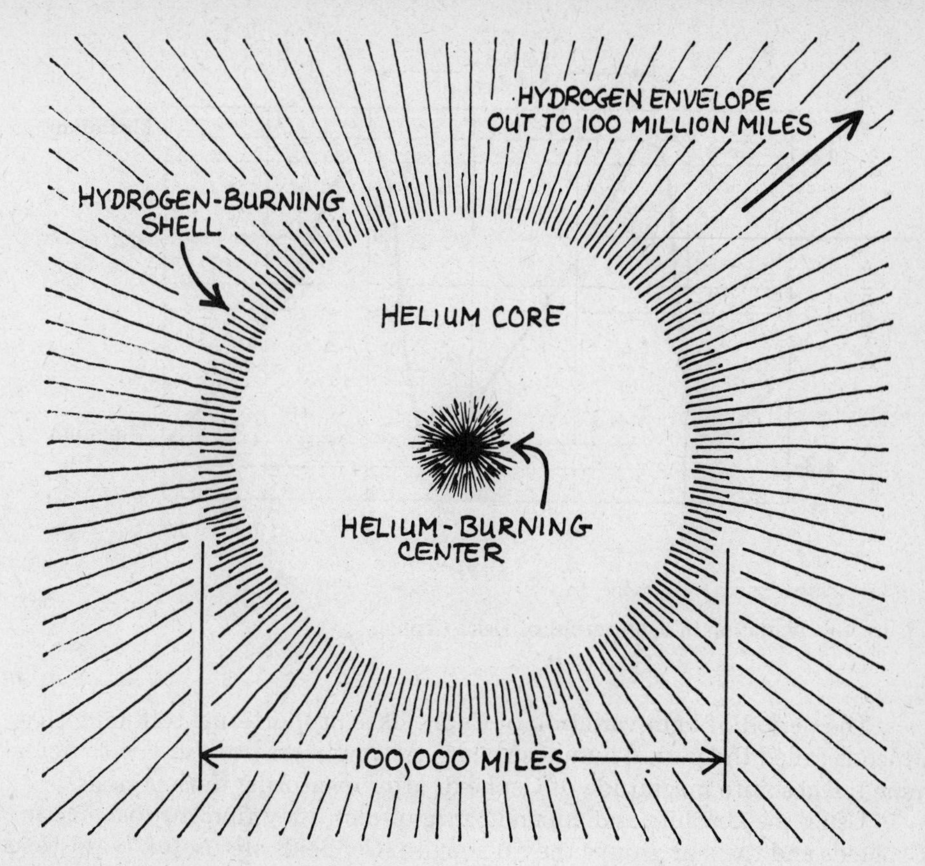

Figure 4.5. Structure of a red giant star.

9. ★ A star probably moves back and forth from the red giant region toward the main sequence several times, in a way not yet fully understood, before it enters the final stages of its life. Most stars probably change from red giants to *pulsating variable stars* before they finally die. That is, they expand and contract and grow bright and fade periodically.

Red supergiants, the largest group of pulsating stars, take between 100 days and 2 years to vary between brightest and faintest. The most famous red supergiant is Mira in the constellation Cetus. Mira was named "The Wonderful" by ancient observers who were stunned to see this star vary periodically from its maximum bright red to its minimum output, where it becomes invisible.

Cepheid variables, large yellow stars, vary their light output in periods of from 1 to 50 days. These stars are rare but important, because they provide a way of measuring distances too great to be measured by parallax.

Figure 4.6 shows how the light output varies for Delta Cephei, the first Cepheid observed and the star for which others were named. You can observe it yourself (see your star maps).

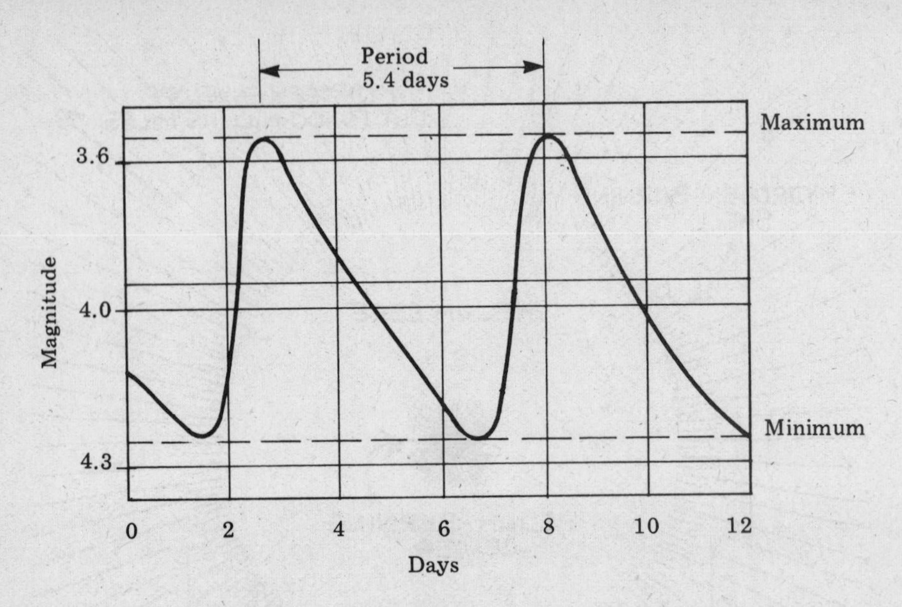

Figure 4.6. Variation in light output of Delta Cephei.

The period of light variation of Cepheids is proportional to luminosity, which is called the period-luminosity law. Astronomers use this law to determine the absolute magnitude of Cepheids after measuring their periods.

Using the absolute and apparent magnitudes, they find the *distance* to Cepheids and the star groups they belong to. Cepheids are useful "yardsticks" out to about 10 million light-years (3 million parsecs). (See frame 3.16.)

Polaris, the North Star, is the nearest Cepheid variable. Its brightness varies between magnitudes 2.5 and 2.6 about every 4 days.

RR Lyrae stars are the second most common variable stars. Almost 4,500 RR Lyrae stars are known in the Milky Way Galaxy. Their light output varies from brightest to dimmest in periods of less than a day. RR Lyrae stars are used to measure the distance to the star clusters they belong to, out to about 600,000 light-years (200,000 parsecs).

What two characteristics of a pulsating variable star change periodically?

---

1. size; 2. luminosity

10. All stars evolve in about the same way, although in different amounts of time, until their cores become mostly accumulated carbon. The last stage in a star's evolution, or the way it finally dies, depends greatly on its *mass.*

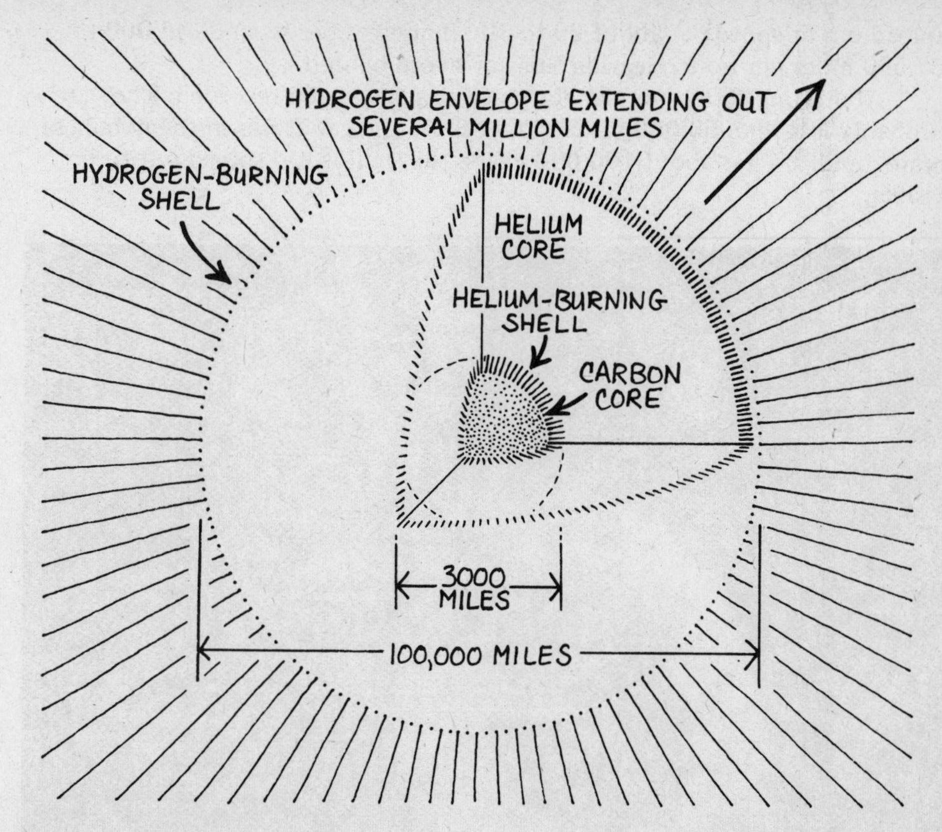

Figure 4.7. The structure of a star's core when inner core of mostly carbon has accumulated.

Small stars, up to about 1.4 times the sun's mass, finally die without a fuss, quietly fading away into the blackness of space. Very massive stars end with a violent explosion, or *supernova*, flaring up brilliantly before giving up life.

What characteristic of a star determines the way it finally dies?

----

its mass

11. When a star of mass like our sun has depleted all of its available helium fuel, it becomes a bloated red giant star for the last time. (At this stage of its life our sun will become so big that it will swallow up Mercury, Venus, Earth, and Mars.)

The star then throws off some of its mass. The star's outermost hydrogen envelope flies off into space. This wispy expanding shell of gas, called a *planetary nebula*, is typically about $\frac{1}{2}$ to 1 light-year across. It continues to spread out at speeds of about 20 to 30 kilometers per second (45,000 to 67,500 miles per hour), leaving the star's core behind.

About one thousand planetary nebulas have been recorded. They are probably less than 50,000 years old, because the gas atoms in the nebula separate rapidly; after about 100,000 years, the shell is too spread out to be visible.

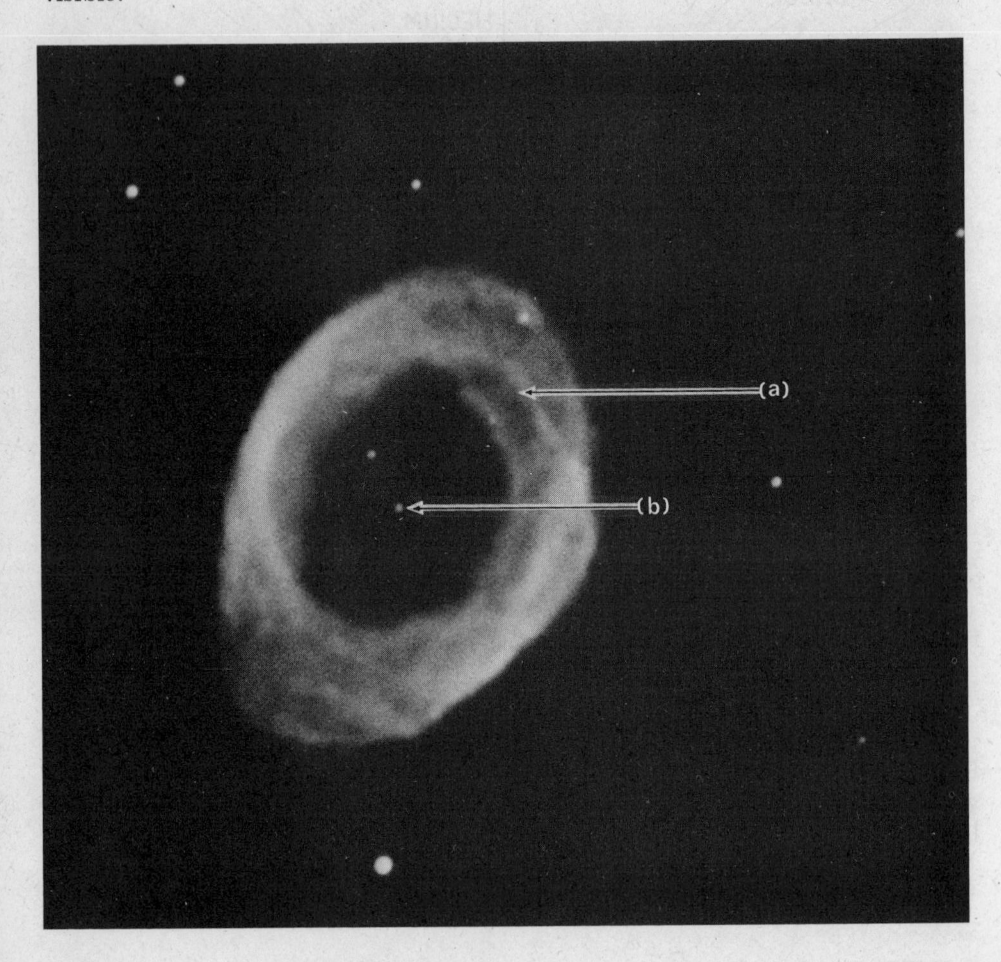

Figure 4.8. Planetary nebula and star core.

Identify the core of the star and the planetary nebula in the photograph.

(a) ______________________ (b) ______________________

- -

(a) planetary nebula; (b) core of the star. The photograph shows the Ring Nebula in Lyra.

12. After it has thrown off its gas envelope, the star remains as a core of carbon surrounded by a shell of burning helium.

A star that has exhausted all of its nuclear fuel can no longer withstand the pull of gravity. It contracts again as gravity pulls matter in toward the center. Gravitational contraction makes the temperature and pressure go up very high, and electrons are stripped off atoms. The star becomes a small, hot, *white dwarf.* It is made mostly of electrons and nuclei. These subatomic particles can be squeezed much closer together than whole atoms can.

Eventually, when the white dwarf star reaches about Earth's size, it cannot contract any further. White dwarf stars of mass like the sun are very dense, because gravity packs all that mass into a star the size of Earth. The force of gravity on such a white dwarf star would be about 350,000 times greater than that on Earth. If you could stand on a white dwarf star, you would weigh 350,000 times more than you do on Earth.

Sometimes at this stage, a *nova*, a brilliant flaring star, is produced. If the white dwarf belongs to a binary system, matter from the companion star may fall onto the white dwarf and fuel the brief, bright flare.

Gradually the white dwarf star cools, turns to dull red, and shines its last energy into space. Then the white dwarf becomes a dead *black dwarf* in the graveyard of space.

What is a white dwarf star? ______________________________________

____

- -

a small dense (dying) star of low luminosity and high surface temperature, typically about the size of Earth but with mass equal to the sun's.

13. Identify each stage of the life of a star like our sun, as labeled sequentially in figure 4.9.

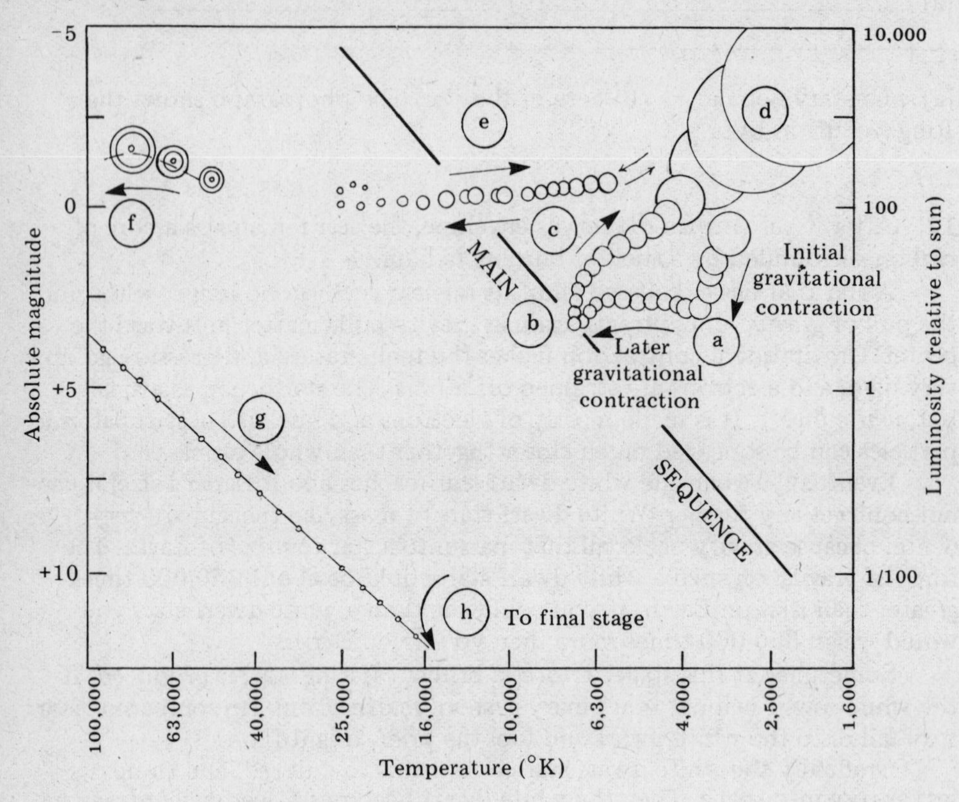

Figure 4.9. The life phases of a star like our sun.

(a) ______________________________

(b) ______________________________

(c) ______________________________

(d) ______________________________

(e) ______________________________

(f) ______________________________

(g) ______________________________

(h) ______________________________

(a) protostar-gravitational contraction of cloud of gas and dust;
(b) stable main sequence star shining by nuclear fusion (converting hydrogen to helium);
(c) evolution to red giant when helium core forms;
(d) red giant, shining by helium burning;

(e) variable star, formation of carbon core;
(f) planetary nebula, hydrogen envelope ejected into space;
(g) white dwarf, mass packed into star about the size of Earth;
(h) dead black dwarf in space.

14. Very massive stars, about four times the sun's mass, die much more spectacularly than stars like our sun. A *supernova* is a gigantic stellar explosion.

A massive star's carbon core contracts because of the force of gravity in the same way that a smaller star's does. But in the more massive star the core temperature continues to rise all the way up to 600 million degrees Kelvin. At that point the carbon core begins to burn. The collapse stops as carbon is converted into magnesium in nuclear fusion reactions.

When the carbon is used up, a new cycle is begun—gravitational contraction, rise in temperature, onset of new nuclear reactions, production of new elements, and a halt in the collapse. Elements heavier than carbon, such as oxygen and silicon, are produced inside the star until the core is mainly iron.

Iron ends these cycles of nuclear fires and collapse, because it does not release energy by nuclear reactions. The doomed star collapses for the last

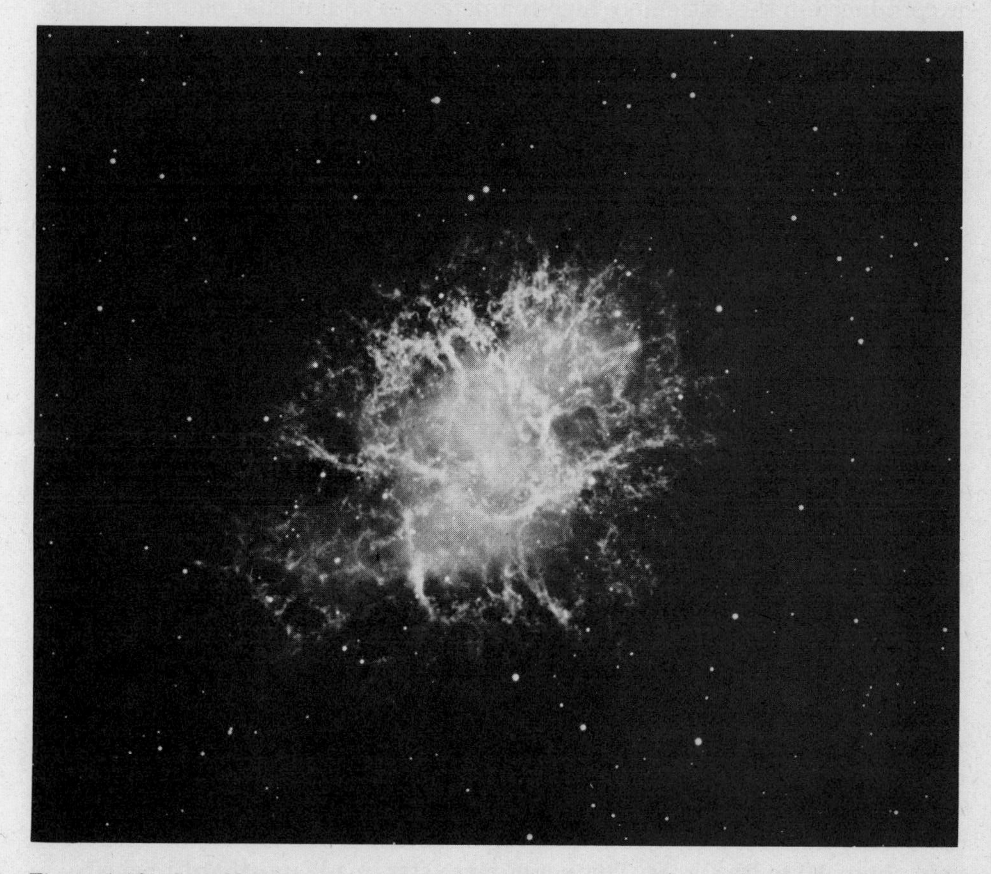

Figure 4.10. Crab Nebula.

time, until it cannot be compressed any more. Then it explodes violently. The light from that supernova can reach a billion times the sun's luminosity.

In 1054 A.D., Chinese and American Indian observers recorded seeing a brilliant new star, a supernova, blaze in the sky even during daylight hours. The *Crab Nebula* (see figure 4.10) in Taurus is a cloud of gas visible today at the site of that supernova. It is now about 10 light-years (3 parsecs) across, with the remnant core of the exploded star still at the center. The Crab Nebula is expanding at about 1,000 miles per second.

What kind of stars die as supernovas? ____________________

- - - - - - - - - - - - - - - - - - - -

very massive stars (about four times the sun's mass)

15. You might say that you are made of star dust. Hydrogen and helium were probably the only elements in the universe when it began. *Elements* such as carbon, oxygen, and nitrogen, essential for life, are made inside the fiery cores of ageing stars. The heaviest elements of all, such as gold and lead, are produced in the extremely high temperatures and intense neutron beams of a supernova explosion.

The supernova explosion sprays all these new elements out into space. They mix with the hydrogen and helium and dust already there. All the material scattered into space by exploding massive stars becomes available again to be used in the formation of new stars and planets. Our sun and Earth were formed about 5 billion years ago from a cloud of hydrogen and helium enriched in this way.

Which do you think are more abundant in the universe, elements lighter than iron or those heavier than iron? Why? ____________________

____

- - - - - - - - - - - - - - - - - - - -

The lighter. Lighter elements have much more time to form. Elements lighter than iron are produced from primordial hydrogen over a long period of time inside the cores of massive stars, while those that are heavier are produced only during the brief interval when the star explodes (supernova) at the end of its life.

16. *Pulsars*, or *pulsa*ting *r*adio *s*tars, were first observed in 1967. Pulsars send sharp, strong bursts of radio waves to Earth with clocklike regularity, at intervals between $\frac{1}{30}$ to 3 seconds long. More than one hundred of these strange objects have been observed so far.

According to theory, when a very massive star explodes, it leaves behind more mass than the sun squeezed tightly together into a superdense ball of only about 10 miles across. This superdense star, made mostly of nuclear particles (called neutrons) in its core, was hypothesized and named a *neutron star* by theorists about 40 years ago.

Theory predicted that a neutron star should exist at the center of the Crab Nebula. A pulsar was found there in 1968; it has since been observed in radio, optical, X-ray, gamma, and infrared wavelengths.

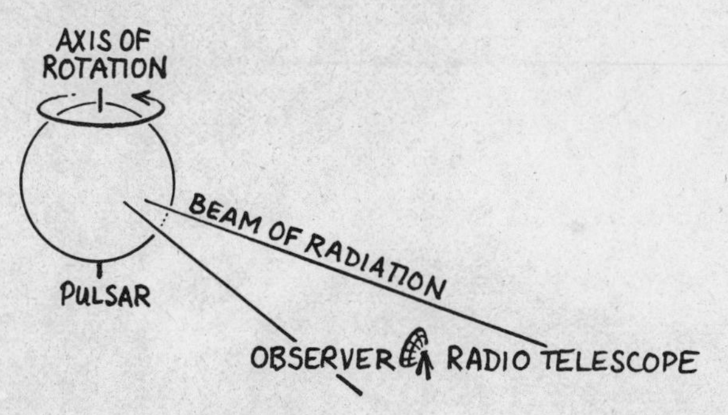

Figure 4.11. A pulsar or neutron star.

A pulsar appears to be a rotating, magnetic neutron star.

How would you expect the force of gravity on the surface of a pulsar, or neutron star, to compare to the force of gravity on Earth? ____________

____

- - - - - - - - - - - - - - - - - -

very much greater on a pulsar. The force of gravity is stronger the closer matter is packed. The pulsar, or neutron star, packs the mass of a star similar to our sun into a ball only 10 miles across, smaller than New York City.

17. A really massive star may continue to collapse after the pulsar stage to become a bizarre object called a *black hole.*

If black holes do exist, they are so superdense that a mass as great as the sun's is packed into a ball less than 2 miles (2.5 kilometers) across. The force of gravity in such a star would be so great that, according to *Einstein's theory of relativity*, it would suck in everything nearby, even light. A black hole can never be seen, because no light, matter, or signal of any kind can ever escape from its gravitational pull.

Cygnus X-1, an X-ray source over 8,000 light-years (2,500 parsecs) distant in Cygnus, is a likely black hole candidate. Cygnus X-1 is an invisible eclipsing binary star (period 5.6 days). Its visible companion is a blue supergiant that shows variations in spectral features from one night to the next. Possibly, when Cygnus X-1 sucks material gravitationally from its visible companion into its rotating disk, the X rays which astronomers detect are emitted.

You will surely hear more about these mysterious black holes in the future. They are one of the most exciting objects being investigated by scientists today.

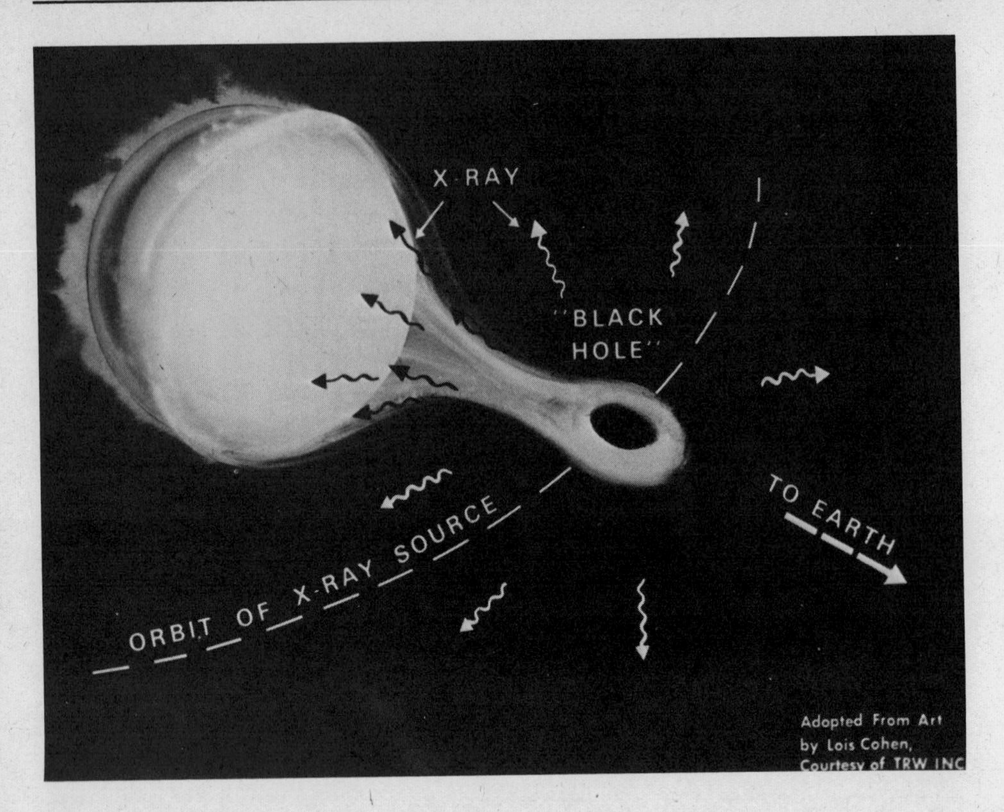

Figure 4.12. Artist's conception of a black hole model.

What do you think would happen if an unlucky spaceship passed close to a black hole in space? ______________________________

____

The strong gravitational pull of the black hole would pull the spaceship in, producing a destructive force that would increase as the ship fell in and that would eventually tear it apart.

SELF-TEST

This self-test is designed to show you whether or not you have mastered the material in Chapter Four. Answer each question to the best of your ability. Correct answers and review instructions are given at the end of the test.

1. Define stellar evolution. ____________________

____

2. How do astronomers check a theory of stellar evolution? ____________________

____

3. List the three main steps in the birth of a star. ____________________

____

4. What is the main source of the energy that a main sequence star shines into space? ____________________

5. For stars of the same initial chemical composition, what property determines the length of time it takes for the stars to evolve?

____

6. Why will the sun stop shining as a main sequence star about 5 billion years from now? ____________________

____

7. List the six main stages in the life cycle of a star like our sun in order from birth to death. ____________________

____

8. Match the descriptions to the main stages in the evolution of very massive stars:

___ (a) formation of elements up to iron	1. black hole
___ (b) gravitational contraction of nebula	2. carbon burning
___ (c) helium burning	3. main sequence
___ (d) possible neutron star	4. protostar
___ (e) possible 2-mile ultradense invisible final stage	5. pulsar
___ (f) shining steadily by conversion of hydrogen to helium	6. red giant
___ (g) violent explosion	7. supernova

9. Why are elements that are lighter than iron, such as hydrogen, helium, carbon, and oxygen, so much more abundant in the universe than are the elements heavier than iron? ____________________

____

10. Match the eight items from the theory of stellar evolution to a real sky object.

___ (a) birthplace of stars
___ (b) black hole candidate
___ (c) blue giant
___ (d) main sequence star
___ (e) neutron star
___ (f) pulsating variable star
___ (g) red giant
___ (h) supernova remnant

1. Betelgeuse (in Orion)
2. Crab Nebula
3. Crab pulsar
4. Cygnus X-1
5. Mira (in Cetus)
6. Orion Nebula
7. Rigel (in Orion)
8. sun

11. What is a black hole? ____________________

____

ANSWERS

Compare your answers to the questions on the self-test with the answers given below. If all of your answers are correct, you are ready to go on to the next chapter. If you missed any questions, review the frames indicated in parentheses following the answer. If you miss several questions, you should probably reread the entire chapter carefully.

1. the changes that take place in stars as they age—the life cycle of stars (frame 1)
2. They predict what changes in luminosity and temperature should take place in stars as they age. Then they compare these theoretical tracks of evolution on H-R diagrams with H-R diagrams for groups of real stars. (frame 1)
3. (1) gravitational contraction of a cloud of gas and dust; (2) rise in interior temperature and pressure; (3) nuclear fusion (frame 3)
4. Nuclear fusion reactions in the core (Hydrogen is converted into helium.) (frames 3, 5)
5. mass (frame 4)

6. The sun will leave the main sequence when all the available hydrogen fuel in its core is used up so that it no longer has an internal energy source. (frame 7)

7. (1) protostar; (2) main sequence star; (3) red giant; (4) variable star; (5) white dwarf; (6) dead black dwarf (frames 3, 5, 6, 7, 8, 9, 12, 13)

8. (a) 2; (b) 4; (c) 6; (d) 5; (e) 1; (f) 3; (g) 7 (frames 3, 5, 6, 7, 14, 16, 17)

9. Hydrogen and some helium were probably the original elements in the universe. The other elements that are lighter than iron are formed inside ageing stars over a period of time. Elements heavier than iron are formed only during the brief time of a supernova. (frame 15)

10. (a) 6; (b) 4; (c) 7; (d) 8; (e) 3; (f) 5; (g) 1; (h) 2 (frames 2, 6, 7, 9, 14, 16, 17)

11. a superdense, gravitationally collapsed mass from which no light, matter, or signal of any kind can ever escape (frame 17)

CHAPTER FIVE
Galaxies

In nature's infinite book of secrecy
A little I can read.

Shakespeare,
Antony and Cleopatra,
Act I, ii, 11

1. ★ A *galaxy* is a group of millions or billions of stars and gas and dust held together in space by the force of gravity.

Our sun and all the visible stars in our sky belong to the *Milky Way Galaxy*. You sometimes see a cloudy band of light across the sky on a very clear dark night. The ancients named it the Milky Way because it looked like a trail of milk spilled in the sky by a goddess who was nursing her baby. That milky band is part of our huge Galaxy.

Try to locate the Milky Way overhead in summer or winter. If possible, use binoculars or a telescope to see that it is really made of many individual bright stars.

The entire Milky Way Galaxy contains over 100 billion stars. Those stars are very far apart from each other in space. On the average, a star's nearest neighbor star is about 5 light-years (30 trillion miles) away.

What is a galaxy? ______________________________

----

an enormous collection of stars and gas and dust held together in space by the force of gravity.

2. **Since we are bound to our sun which is located inside the huge Milky Way Galaxy, we cannot photograph our own Galaxy from the outside.** (To try to do so would be similar to trying to take an aerial view of Washington, D.C., from the base of the Washington Monument!) Instead we use photo-

Figure 5.1. A star cloud in the region of Sagittarius.

graphs of distant galaxies to help us picture what our own Galaxy must look like from outside.

If you could go far out into space and look down on our Galaxy, you would see a brilliant spiral pinwheel almost 100,000 light-years (30,000 parsecs) across. Our Earth, travelling around our sun, is located out in one

Figure 5.2. Galaxy M74. The arrow shows where our sun would be located if this were our Galaxy.

of the spiral arms. (See figure 5.2. The arrow shows where our sun would be located if this were our Galaxy.)

If you could look at the Milky Way Galaxy from the side, it would look something like a phonograph disk with a swollen label. The thickness of the center bulge is about 10,000 light-years (3,000 parsecs). The thickness of the disk is about 1,000–2,000 light-years (300–500 parsecs). Our sun is about 30,000 light-years (10,000 ± 800 parsecs) away from the center. (See arrow in figure 5.3.)

The whole Milky Way Galaxy is turning around in space. This fact is deduced from the Doppler shift of radiation from the spiral arms. Our sun, with its family of planets, is racing around the center of our Galaxy at about

Figure 5.3. Galaxy NGC4565, viewed edge-on.

563,000 miles per hour (250 kilometers per second). Even at that incredible speed, our solar system requires roughly 200 million years to complete just one trip around the center of our huge Galaxy.

Our Galaxy appears to be hurtling through space in the direction of the constellation Hydra at a speed of over 1,000,000 miles per hour.

Roughly how big is the diameter of our Galaxy (a) in light-years?

_______________ (b) in miles? _______________

- - - - - - - - - - - - - - - - - - - -

(a) 100,000 light-years; (b) 6 hundred thousand trillion miles

Solution: Multiply (100,000 light-years) x (6 trillion miles per light-year)

3. Our Galaxy is called a *spiral* type because of its shape. In spiral galaxies most of the stars are concentrated in the central *nucleus* and in the *spiral arms* that radiate out away from the center.

While some individual stars travel through the Galaxy alone, many stars move together in the disk as groups called *open*, or *galactic*, *clusters* (figure

Figure 5.4. Pleiades star cluster (M45, NGC1432) in Taurus, showing stars and surrounding nebulosity.

5.4). More than one thousand open (galactic) clusters containing from ten to one thousand stars each have been observed. They are strongly concentrated in the spiral arms.

A small fraction of the stars are in an estimated five hundred *globular clusters* in a *halo* around the disk (figure 5.5). About 125 globular clusters, containing between 10,000 and a million stars each, have been detected.

Figure 5.5 Globular star cluster (M92, NGC 6341) in Hercules.

The large-scale features of a *normal galaxy* such as ours, are a composite of the properties of the individual stars it contains. The major part of our Galaxy's mass is concentrated in stars like our sun, whose mass is 1.99×10^{33} grams. What would you expect the *mass* of the whole Milky Way Galaxy to be? Hint: Use the approximate number of stars in our Galaxy from frame 5.1.

---

over 100 billion times the sun's mass, or more than 2×10^{44} grams

Solution: Multiply (100 billion stars) $\times$ (1.99×10^{33} grams) = about 2×10^{44} grams. (Note: The mass of the Milky Way Galaxy is 140 billion times the sun's mass.)

4. The space between the stars is virtually empty. The *interstellar medium*, the matter and radiation between the stars, is much less dense than the air in the best vacuum produced on Earth.

Interstellar matter is about 99 percent gas (mainly hydrogen) and 1 percent very tiny solid particles called *interstellar dust.* Interstellar matter is particularly important, because it is the raw material for new stars and planets. In our Galaxy most of the interstellar gas and dust is concentrated in the spiral arms, and that is where the newest stars are located.

Great excitement has arisen over recent discoveries of a large variety of *molecules* in the interstellar medium. *Water vapor* and common *organic molecules* are among the more than fifty molecules that have already been detected. These are key components of all known life on Earth.

Many intriguing questions about the origin of life in the universe have been raised by the discovery of these interstellar molecules. Watch for more on this topic in current publications as new discoveries occur.

Why is it important in the theory of stellar evolution to know what interstellar matter consists of in any epoch? ______________________________

____

- - - - - - - - - - - - - - - - - - -

Interstellar matter is the raw material for new stars and planets.

5. About 12 billion years ago, our Galaxy formed out of a turbulent cosmic cloud made up mostly of hydrogen and helium that was present when our universe began.

What would you expect (a) the oldest and (b) the youngest stars in our Galaxy to be made of? Explain your answer. ______________________________

____

- - - - - - - - - - - - - - - - - - -

(a) hydrogen and helium, the elements present as raw materials at the time our Galaxy was new

(b) hydrogen, helium, and the other ninety naturally occurring elements. The interstellar medium is the raw material for new stars. Originally hydrogen and helium, it has been enriched by elements manufactured inside stars and sprayed into space by supernovas.

6. ★ Some regions of the Milky Way Galaxy have a concentration of gas and dust called a *nebula*, from the Latin word for cloud.

A bright *emission nebula* is a cloud that glows by absorbing and then reemitting starlight from very hot, young stars embedded in it. The Orion Nebula (frame 4.2) is an example you can observe.

A *dark nebula* is a relatively dense concentration of interstellar matter whose dust absorbs or scatters starlight and hides stars that are behind it from our view.

Figure 5.6. Horsehead Nebula (IC434) in Orion.

Some nebulas are given fanciful names according to their appearances. Refer to figure 5.6. What is the "horse's head" actually made of?

---

relatively dense concentrations of interstellar dust

---

7. We cannot look very far into our Milky Way Galaxy even with the biggest optical telescopes because dust clouds block our view. Radio astronomers use *radio waves* that can pass through these clouds to give us a picture of our Galaxy (frame 2.26).

The spiral structure of our Galaxy is mapped by detecting radio waves of *21-centimeter* wavelength. This 21-centimeter radiation is emitted by *hydrogen atoms.* It is strongest from the regions which have the biggest concentration of hydrogen atoms—the spiral arms.

The 21-centimeter radiation does not reveal exceptionally dense concentrations of hydrogen in the spiral arms. In these regions hydrogen atoms join together to form hydrogen molecules. Molecular hydrogen does not emit the 21-centimeter line. Radio astronomers may map the densest gas concentrations by looking at a strong carbon monoxide emission line.

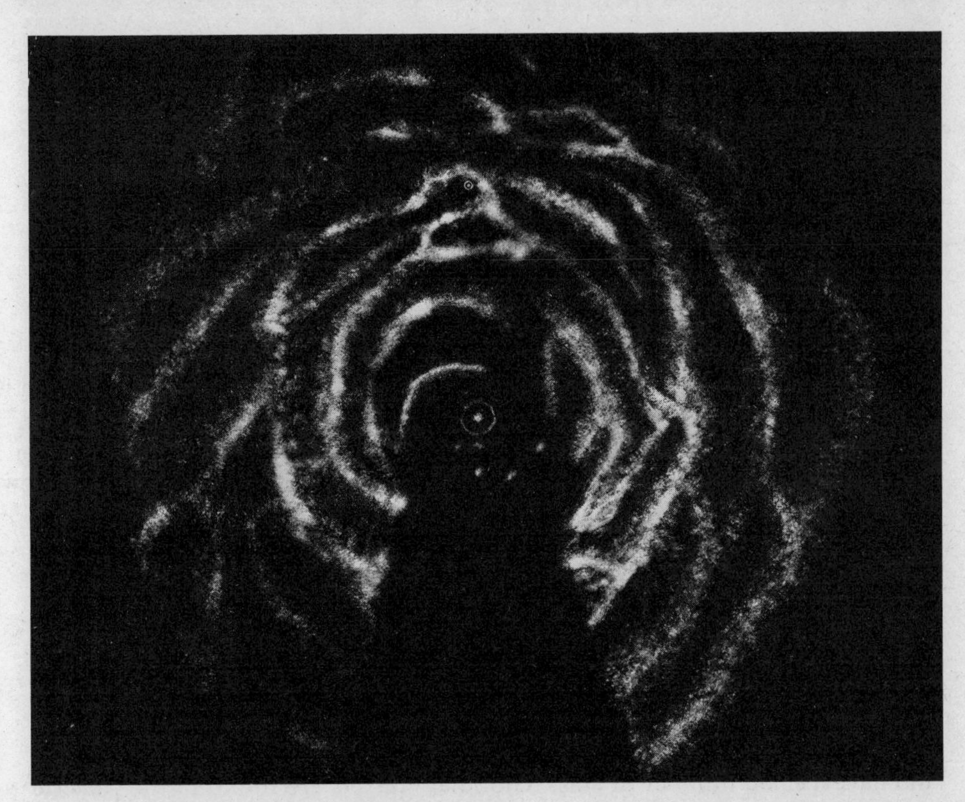

Figure 5.7. A radio map of the Galaxy produced by Leiden Observatory from observations of the 21-centimeter line. Large circle locates galactic center and small circle locates our solar system.

Today astronomers are studying galactic objects using infrared and high energy radiations. New data continue to modify our picture of the Galaxy. Recent observations include the Galactic center in the infrared, stellar coronas in the ultraviolet, and X-ray and gamma ray bursters.

What is particularly interesting about regions of relatively dense gas concentrations in our Galaxy? ______________________

— — — — — — — — — — — — — — — — — — — —

New stars are forming in these regions.

8. ★ *Star clusters* are groups of stars that stay together because of their mutual gravitational attraction. They are important to astronomers because all the stars in a cluster are about the same age.

Star clusters form as a huge cloud of gas condenses into many stars rather than into just one single star. We see evidence of the cluster origin of stars today, because a high proportion of observed stars are binary stars (frame 3.21) or members of *multiple star systems* (made up of more than two stars).

Table 5.1. Some properties of star clusters

	Open (Galactic) Clusters	Globular Clusters
Location	galactic disk	galactic halo
Age	relatively young	old
Number of stars	up to 1,000	up to 1 million
Color of brightest stars	blue or red	red
Composition	heavy elements present; metal-rich	heavy elements not present; metal-poor

The Pleiades star cluster (figure 5.4) is an example of the open, or galactic, cluster; figure 5.5 shows a globular cluster in Hercules. You can observe both (see Appendix 4). Refer to Table 5.1. List three differences between the open (galactic) clusters and the globular clusters found in our Galaxy. Note: Review frame 3.

- - - - - - - - - - - - - - - - - - -

Open (galactic) clusters are found in the galactic disk, are relatively young, and have a small number of stars. Globular clusters are found in the galactic halo, are relatively old, and have a large number of stars.

9. Star clusters provide the best data for verifying a theory of stellar evolution.

First, H-R diagrams predicted by theory for stars of different ages are drawn. Then H-R diagrams for observed star clusters are drawn. The theoretical and observed diagrams are compared to verify or disprove the theory.

Figure 5.8 is a representation of predicted evolutionary tracks computed from theory. All stars start on the main sequence when they are born. The most massive stars are located at the top of the main sequence, and the least massive are at the bottom. All stars evolve away from the main sequence as they age. Massive stars evolve fastest, so the higher the turnoff point, the younger the star cluster.

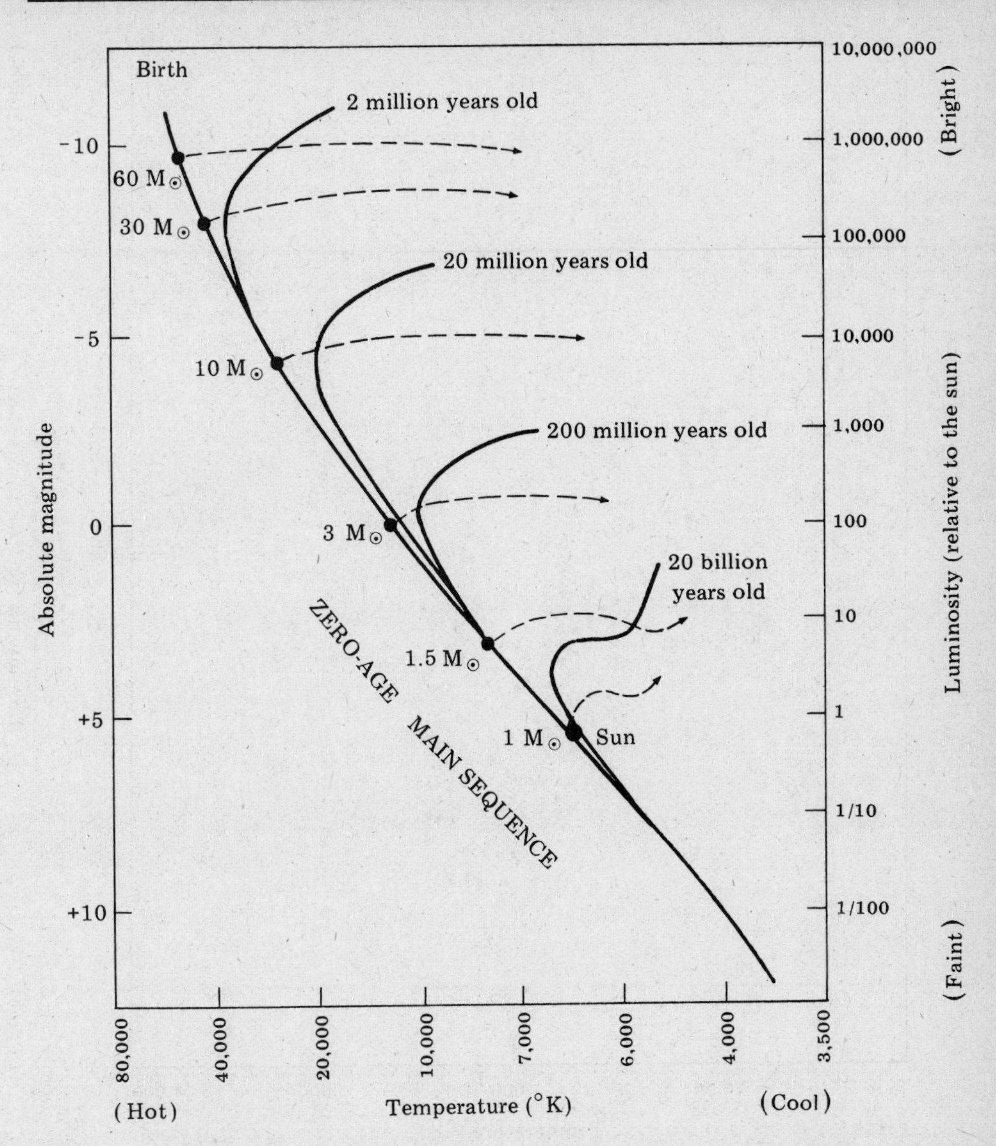

Figure 5.8. Solid lines give positions for stars in a cluster, at various times following birth of the cluster.

Dashed lines show tracks of evolution of individual stars with masses shown.

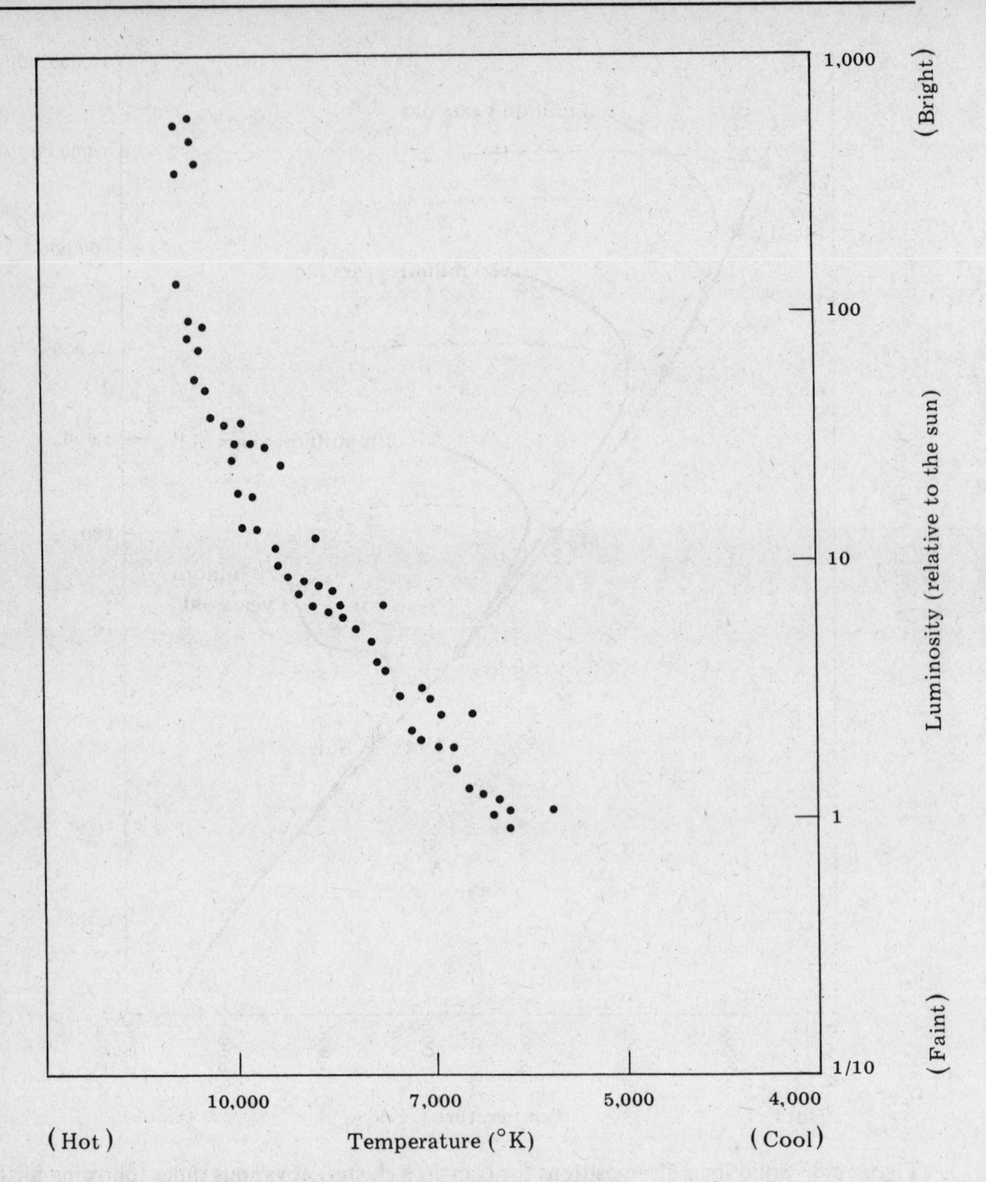

Figure 5.9. H-R diagram of Pleiades open (galactic) cluster.

Compare the H-R diagrams for the Pleiades cluster (figure 5.9) and the M3 globular cluster (figure 5.10) with the theoretical tracks in figure 5.8. State which is (a) a young cluster ________________; (b) an old cluster ________________. Explain your reasoning. ________________

____

- - - - - - - - - - - - - - - - - -

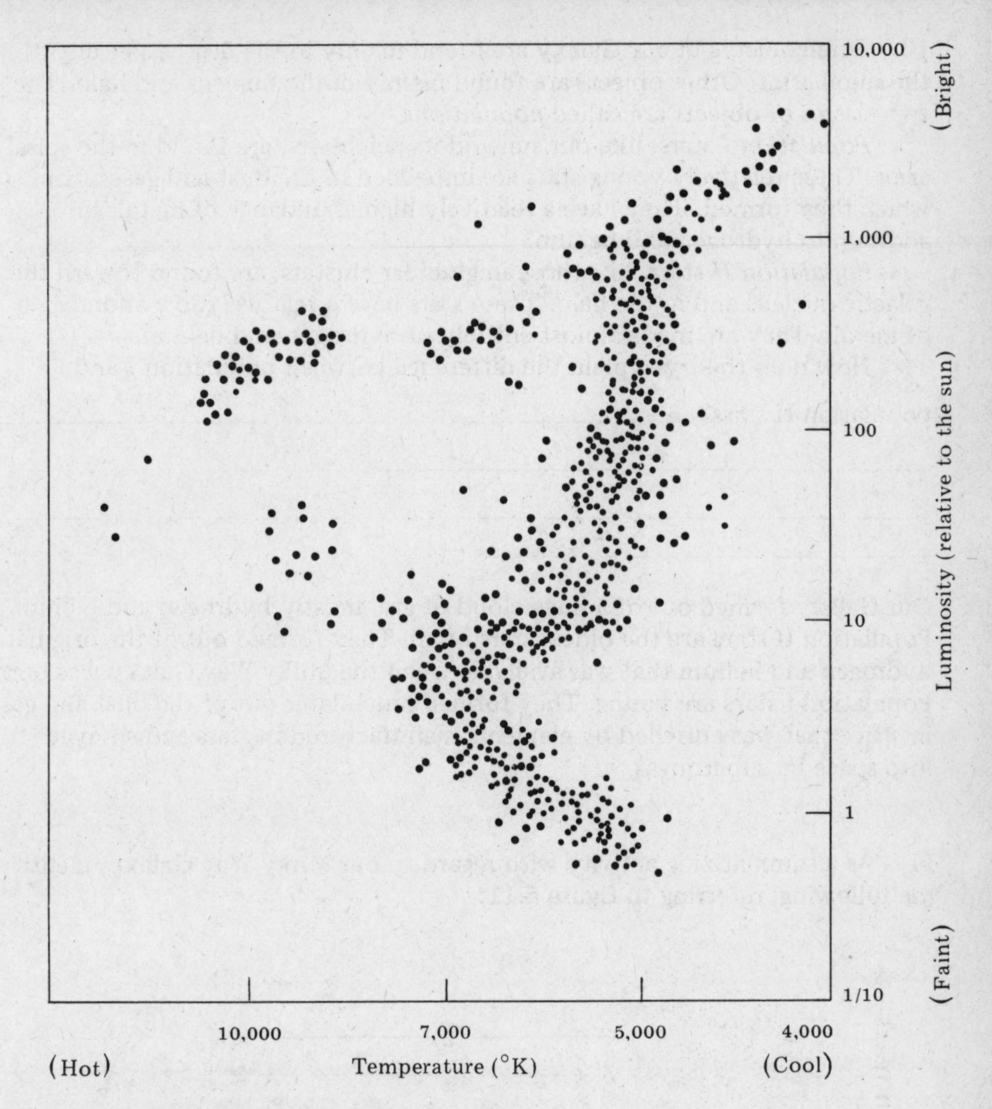

Figure 5.10. H-R diagram for globular cluster M3.

(a) Pleiades cluster is young. Most of its stars, even the massive short-lived ones are still on the main sequence. (Pleiades cluster was born only about 70 million years ago.)

(b) M3 is old. Hardly any stars appear on the upper half of the main sequence, and many stars have moved to the right into the red giant region. (M3 is about 10 billion years old.)

10. Some objects in our Galaxy are found mainly in the disk, especially in the spiral arms. Other objects are found mainly in the nucleus and halo. The two classes of objects are called *populations.*

Population I stars, like our sun and its neighbors, are found in the spiral arms. These relatively young stars are imbedded in the dust and gases from which they formed. They have a relatively high abundance of metals in addition to hydrogen and helium.

Population II stars, like those in globular clusters, are found toward the galactic nucleus and in the halo. These stars have a relatively low abundance of metals. They are made almost entirely of hydrogen and helium.

How does theory explain the difference between population I and population II stars? ____

____

____

Our Galaxy formed out of a huge cloud of gas, mostly hydrogen and helium. Population II stars are the oldest stars of all. They formed out of the original hydrogen and helium that was available when the Milky Way Galaxy was born. Population I stars are young. They formed much later out of the dust and gas in space that was enriched by elements manufactured in stars and sprayed into space by supernovas.

11. As a summarizing activity with regard to our Milky Way Galaxy, identify the following, referring to figure 5.11:

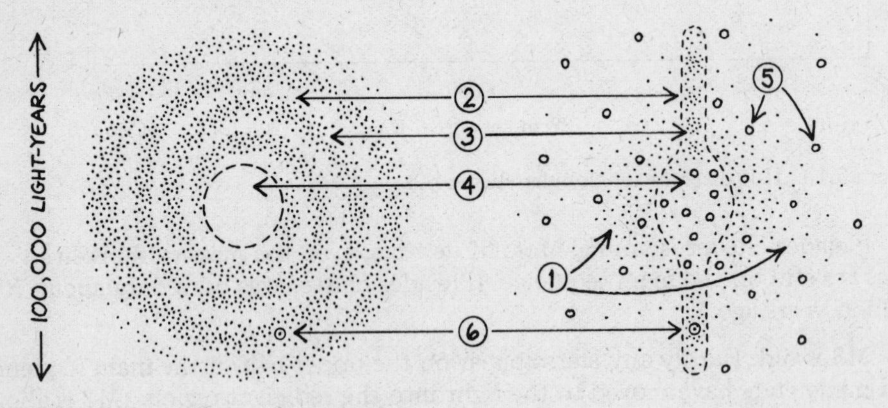

Figure 5.11. Two views of the Milky Way.

___ (a) disk

___ (b) halo

___ (c) spiral arm

___ (d) nucleus

___ (e) position of sun and Earth

___ (f) location of globular clusters.

(a) 2; (b) 1; (c) 3; (d) 4; (e) 6; (f) 5

12. ★ Until a half century ago our Galaxy was the only one recognized. Then Edwin Hubble proved, in 1924, that some of the fuzzy "nebulas" previously observed were really distant *galaxies.*

Today we believe that the universe is full of galaxies, perhaps 100 billion of them, having typically 100 billion stars each. Most bright galaxies are referred to by their New General Catalogue (NGC) or Messier (M) number (see Appendix 4).

The two small, irregularly shaped *Magellanic Clouds* are nearby satellite galaxies of our Milky Way, held by the force of gravity at a distance of about 150,000 light-years (46,000 pc and 63,000 pc). They are visible to the naked eye from the southern hemisphere and were first noted by the explorer Ferdinand Magellan on his trip around the world in the sixteenth century.

The *Andromeda Galaxy* (M31, NGC224) is the closest galaxy that is similar to ours in size and shape. It is about 2 million light-years away. It is the most distant object you can see in space with your naked eye. In the fall, look for a fuzzy patch of light (shown as O Galaxy on your star map) in the Andromeda constellation. The Andromeda Galaxy is even bigger than our Galaxy and contains billions of individual stars.

Figure 5.12. Andromeda Galaxy photographed through a 48-inch telescope.

To get an idea of what "close" means for galaxies, estimate how many Milky Way Galaxies you could line up next to each other between us and our neighbor, Andromeda Galaxy. ____________________

- - - - - - - - - - - - - - - - - -

Twenty.

$$\text{Solution: } \frac{\text{distance to Andromeda Galaxy}}{\text{diameter of Milky Way Galaxy}} = \frac{2{,}000{,}000 \text{ LY}}{100{,}000 \text{ LY}} = 20$$

13. Galaxies come in several different shapes and sizes. They were first classified into groups according to their *structure* by Edwin *Hubble* in 1926.

Elliptical galaxies, designated *E*, are egg shaped. They range from nearly perfect spheres, *E0*, to the flattest, *E7*. Elliptical galaxies do not have any conspicuous gas and dust to make new stars. They seem to contain only old stars.

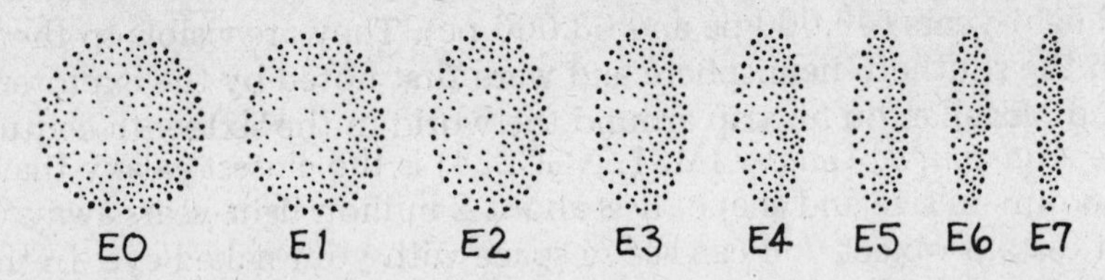

Figure 5.13. Elliptical galaxies.

Spiral galaxies are divided into two general types. *Normal spiral galaxies*, *S*, like our Milky Way, have a flattened disk, a nucleus, a halo, and spiral arms radiating from the center. They are labelled *Sa*, *Sb*, or *Sc*, according to how tightly wound the spiral arms are. Spiral galaxies have large amounts of gas and dust in the disk and contain young, middle-aged, and old stars.

Figure 5.14. Normal spiral galaxies.

Barred spiral galaxies, *SB*, look like normal spiral galaxies except that the spiral arms unwind from the ends of a bar-shaped concentration of material.

Irregular galaxies, *Ir*, have no regular geometric shape. They usually contain gas and dust, mostly bright young stars, and some old stars.

Figure 5.15. Barred spiral galaxies.

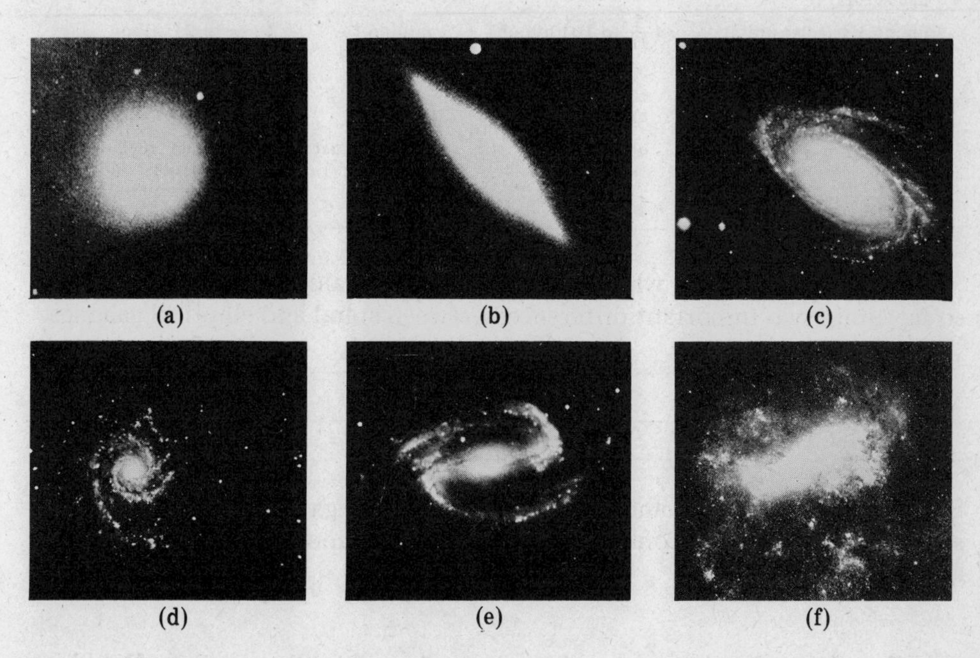

Figure 5.16. Galaxies of various classifications.

Identify each of the galaxies in figure 5.16 according to Hubble classification.

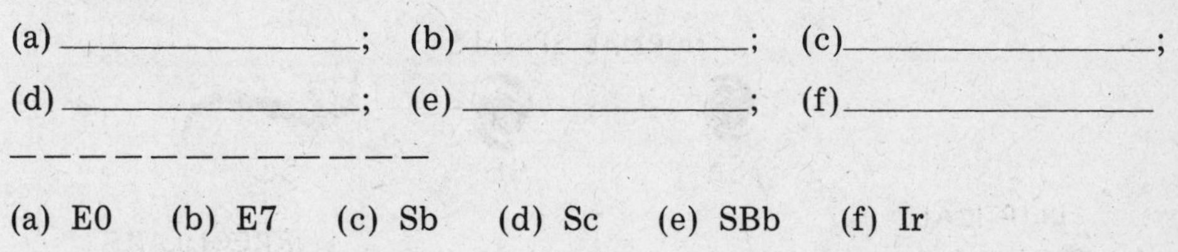

(a) ______________; (b) ______________; (c) ______________;

(d) ______________; (e) ______________; (f) ______________

(a) E0 (b) E7 (c) Sb (d) Sc (e) SBb (f) Ir

14. Hubble began the systematic study of distant galaxies using the 100-inch Mount Wilson telescope, the world's largest from 1918 to 1938. Today astronomers are gathering much more data on these distant star systems using radio, infrared, and X-ray detectors in addition to the giant optical telescopes.

Table 5.2. Rough values of galactic data

Values	Spirals	Ellipticals	Irregulars
Mass (solar masses)	$\left(1 \text{ to } 400\right)$ billion	1 million to 10 trillion	$\left(\frac{1}{10} \text{ to } 30\right)$ billion
Diameter (thousands of light-years)	20 to 150	2 to 500 (?)	5 to 30
Luminosity (solar units)	$\left(\frac{1}{10} \text{ to } 10\right)$ billion	1 million to 100 billion	$\left(\frac{1}{100} \text{ to } 2\right)$ billion
Absolute visual magnitude	−15 to −21	−9 to −23	−13 to −18
Population content of stars	old and young	old	old and young
Interstellar matter	both gas and dust	almost no dust; little gas	much gas; much dust in some; less and sometimes no dust in others

Refer to Table 5.2, which summarizes rough values of the data collected so far. State two important differences between spiral and elliptical galaxies.

____

____

- - - - - - - - - - - - - - - - - -

Spirals contain both old and young stars; they have gas and dust between the stars. Ellipticals contain only old stars; they have almost no interstellar gas and dust.

15. The reason for the varying shapes of galaxies is still a mystery. Hubble arranged the different-shaped galaxies along a "tuning fork" (see figure 5.17). He suggested that a newborn galaxy starts out elliptical, changes to spiral as it ages, and finally dies as an irregular galaxy.

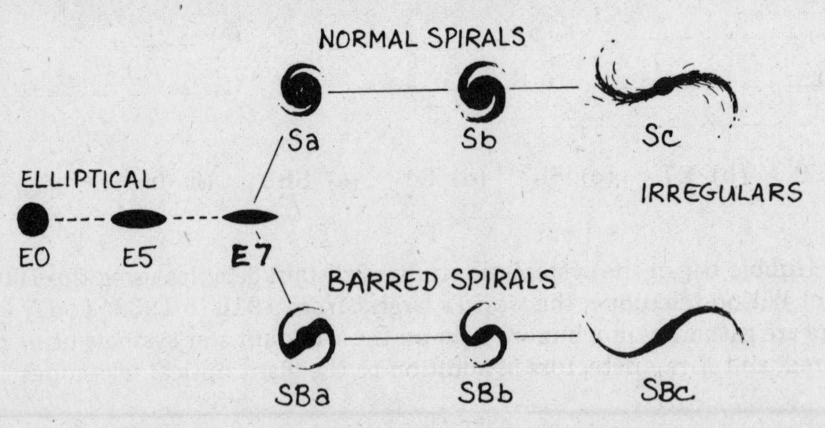

Figure 5.17. The Hubble "tuning fork" diagram of galactic shapes.

A theory suggesting that a galaxy evolves, or goes through a life cycle, in the opposite direction, starting out as an irregular, collapsing to a spiral, and finally winding tighter into an elliptical, has also been proposed.

What observed data (see Table 5.2) indicate that the different shapes of galaxies do not represent stages of evolution in the life cycle of a galaxy?

All three types of galaxies contain old stars. That fact indicates that spiral and irregular galaxies are just as old as elliptical galaxies and could not be the end stage of a galaxy's life. Nor could elliptical galaxies be the first stage of a life cycle, as suggested by Hubble, because they do not have dust and gas which are necessary for the birth of new stars seen in spiral galaxies.

Current theory says that elliptical and spiral galaxies are fundamentally different from each other and do not change into each other as time goes by. Watch for more news on this fascinating research topic in the next few years.

16. Photographic surveys of the sky show that most galaxies belong to groups, called *clusters of galaxies.* These clusters contain from several to thousands of galaxies. They are held together by the force of gravity as they move around a common center of mass.

Our Milky Way Galaxy belongs to a typical small cluster with over twenty members, called the *Local Group.* "Local" for a group of galaxies means they are within a region 3 million light-years across. Three of these galaxies—our Milky Way, Andromeda, and M33 in Triangulum—are spirals. The others are ellipticals, or irregulars.

A rich cluster contains at least one thousand members. The nearest rich cluster of galaxies is about 62 million light-years (19 million parsecs) away in the direction of the constellation Virgo. It contains about twenty-five hundred galaxies.

Nearly three thousand rich clusters of galaxies have been catalogued by George O. Abell. The most distant are probably as far as 4 billion light-years away. Equal numbers of clusters of galaxies seem to be located in all directions in space overall so that the universe looks homogeneous on the large scale.

A cluster of clusters of galaxies is called a *supercluster*. Our Milky Way Galaxy is a part of the *Virgo Supercluster*. Superclusters are the largest gravitationally bound systems found in the universe so far.

What is the largest known type of structure in the universe?

---

supercluster of galaxies

17. Some galaxies, called *peculiar galaxies*, have strange forms and unusual characteristics.

Exploding galaxies show evidence of a violent explosion. They have streamers of hydrogen gas racing outwards to 12,000 light-years from the center at speeds up to 600 miles per second. These galaxies emit exceptionally large amounts of energy in radio, infrared, or X-ray wavelengths. This energy cannot be produced by normal nuclear fusion reactions in stars in normal galaxies.

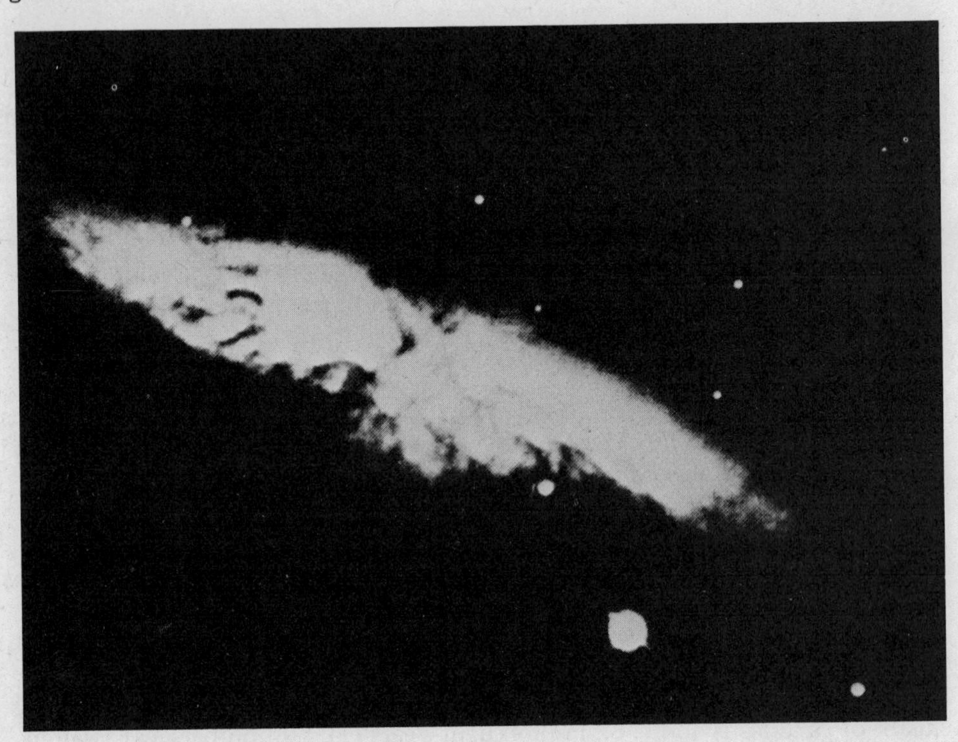

Figure 5.18. M82, an exploding galaxy.

Several hundred apparent collisions between galaxies and cases of galaxies passing close to each other have been photographed. When two galaxies collide, they probably pass through each other. The clouds of gas and dust in *colliding galaxies* would be much more dense, increasing the likelihood of the formation of new stars.

Figure 5.19. NGC2685, an apparent collision of two galaxies.

What do you think might happen to life on Earth if our Galaxy collided with another galaxy? Explain your answer. ______________________________

____

Probably nothing. The stars and their planets are separated by such vast distances inside galaxies that two galaxies can pass through each other without their stars ever coming into contact with each other. (No star collisions have ever been observed.)

18. *Radio galaxies* emit exceptionally large amounts of energy in the radio wavelength region, sometimes as much as a million supernovas exploding at once. The radiograph of a typical radio galaxy shows two large patches emitting radio waves on opposite sides of a visible galaxy located between them.

Astronomers are puzzled by the question of where all the radio energy comes from. The radio energy usually looks like *synchrotron radiation,* or radiation produced by electrons spiraling around at nearly the speed of light in a magnetic field. Perhaps violent events inside the galaxy cause electrons to circulate in a magnetic field that surrounds the galaxy and emit unusual amounts of energy.

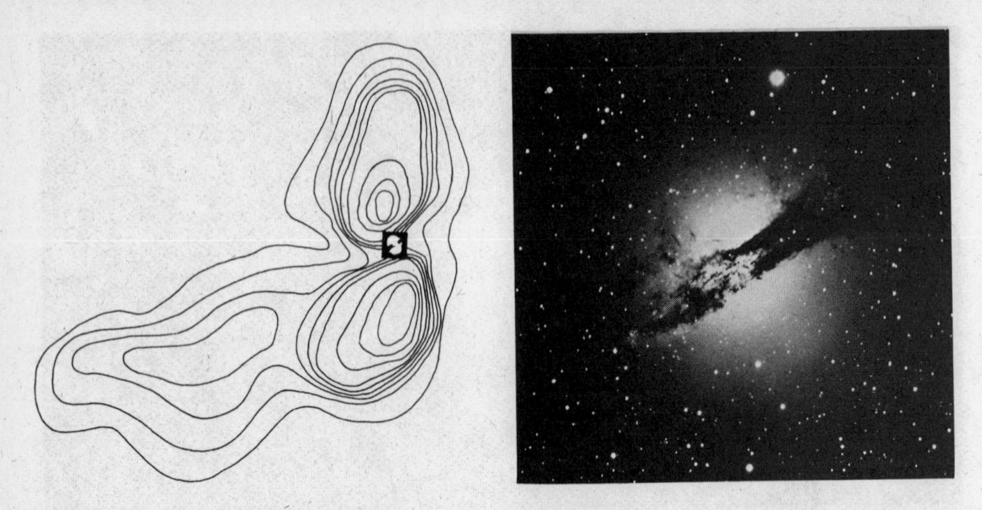

Figure 5.20. The radio image of NGC5128, or Centaurus A.

What is a possible explanation of the Centaurus A radio source shown in figure 5.20? ______________________________

____

- - - - - - - - - - - - - - - - - - -

A violent event, perhaps an explosion, took place at the site of the visible galaxy long ago, and two huge clouds of matter are still expanding outward from that site, emitting powerful radio signals.

19. *Seyfert galaxies* look like spirals in long-exposure photographs but are very different. A Seyfert nucleus, typically only about 10 light-years across, shines one hundred times more brilliantly, mostly in infrared wavelengths, than a normal galaxy the size of our Milky Way. Astronomers are unsure of the explanation for the exceptional energy output, but they are eager to find one.

How is a Seyfert galaxy different from a normal spiral galaxy?

____

- - - - - - - - - - - - - - - - - - -

Seyfert galaxies have a small, brilliant nucleus that greatly outshines all normal galaxies.

20. Among the most intriguing celestial objects of all are the puzzling *quasars.*

Quasars, or *qua*si-*s*tellar *r*adio *s*ources, look like faint stars on ordinary photographs. Quasars are small for celestial objects, typically about 1 light-

Figure 5.21. Quasar 3C295 in Virgo.

year across (not much bigger than our solar system), but they are believed to shine brighter than a hundred normal galaxies full of billions of suns. They have been observed at infrared and high energy wavelengths, too.

The light from quasars is highly shifted toward the red end of the spectrum. Most astronomers interpret this quality as a Doppler shift, meaning the quasars are racing *away* from us at speeds of as much as 91 percent the speed of light. If their assumptions are accurate, quasars are the most distant and powerful objects we have ever photographed. Quasar 0Q172 in the constellation Boötes has the largest redshift measured. When this quasar emits ultraviolet light, it is received as red light on Earth. If indeed this effect is a Doppler redshift, Quasar 0Q172 is at the edge of the known universe, racing away at a fantastic speed of over 600 million miles per hour.

Twin or multiple images of what seem to be the same quasars support the view that quasars are located at cosmological distances. According to Einstein's theory of general relativity, starlight passing near a massive body is deflected. A galaxy much closer to us than a particular quasar could be a *gravitational lens* that produces multiple images of it.

Some astronomers disagree. They deduce from Einstein's theory of relativity that an enormous gravitational force may cause the quasars' extraordinary redshift (a *gravitational redshift*). If so, then the quasars may not, after all, be so far away from our galaxy or so powerful.

Different hypotheses have been proposed to explain the stupendous energy output of these cosmic powerhouses, assuming they are really extremely distant. Quasars may belong to a family of galaxies, since they resemble Seyfert galaxies in brightness, size, and spectrum. They may get energy from collisions of material particles with exotic *antimatter.* They may even have a new source of energy unknown in physics today. Watch for the explanation of the mysterious quasars in the years ahead.

What is so mysterious about the quasars? ______________________

____

----

the source of the stupendous energy output they must have if they are really so extremely far away as believed

SELF-TEST

This self-test is designed to show you whether or not you have mastered the material in Chapter Five. Answer each question to the best of your ability. Correct answers and review instructions are given at the end of the test.

1. Define a galaxy. ______________________________
____

2. Arrange the following in order of increasing size: star, planet, galaxy, cluster of galaxies, open cluster, supercluster, solar system. ____________
____

3. Sketch an edge-on view of the Milky Way Galaxy, and label the (a) size of the diameter; (b) disk; (c) nucleus; (d) spiral arm; (e) halo; (f) position of sun and Earth; (g) location of globular clusters.

4. Which of the following have been identified in the interstellar medium: hydrogen gas, radiation, bacteria, tiny solid dust particles, viruses, water vapor, spirits, gases of elements heavier than hydrogen, organic molecules, algae?

5. Why is it important in the theory of stellar evolution to know what interstellar matter consists of in any epoch? ______________________
____

6. Refer to figure 5.6. Is the space behind the "horse's head" really empty of stars? ______ Explain. ______________________________

7. Why are the radio waves of 21-centimeter wavelength emitted by hydrogen atoms more useful than visible light in mapping the structure of our Milky Way Galaxy? ______________________________
____

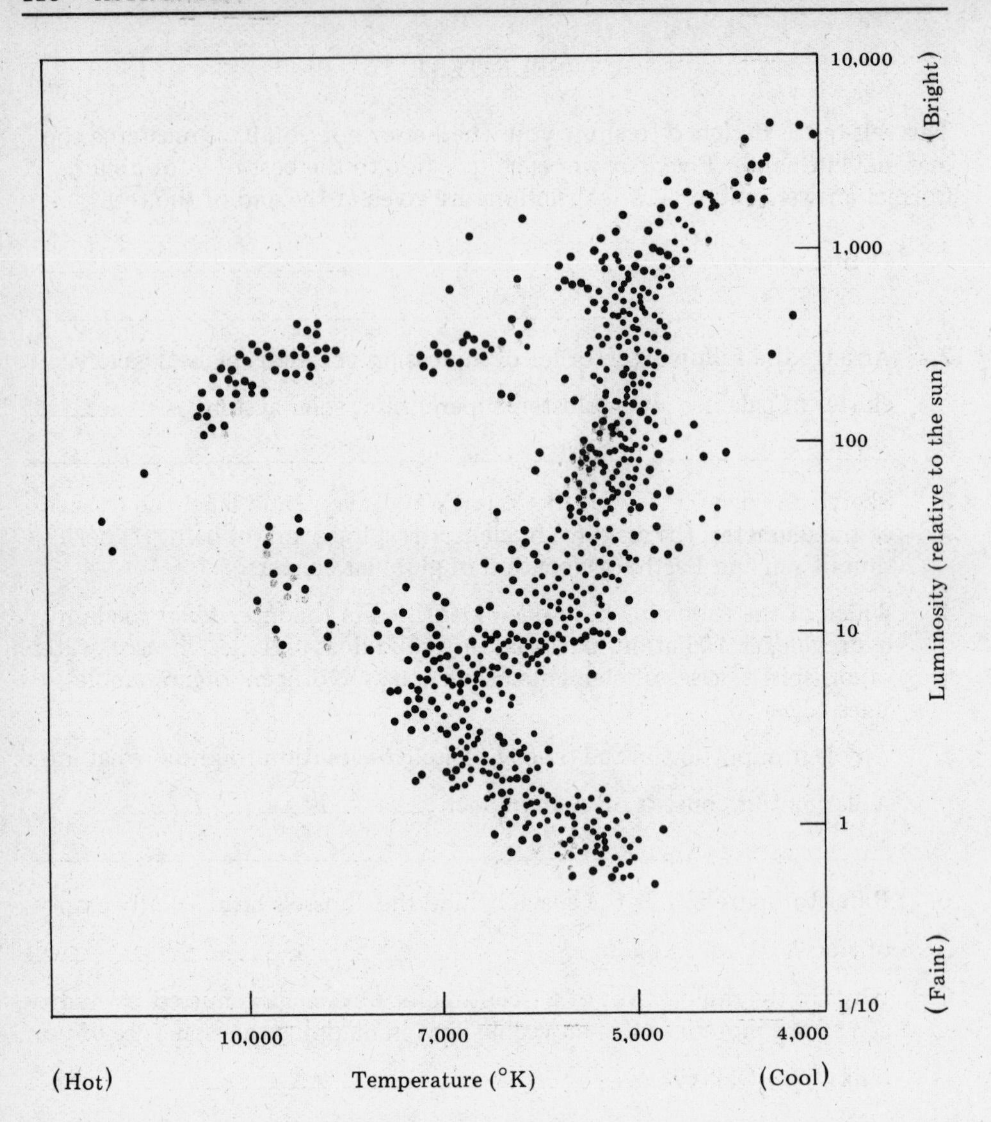

Figure 5.22. H-R diagram ("Cluster 1").

8. Refer to the H-R diagrams of star clusters 1 and 2 (figures 5.22 and 5.23).

   ___ (a) Which cluster is older?

   ___ (b) Which cluster would have population I stars?

   ___ (c) Which cluster would have stars with a relatively high abundance of metals?

   ___ (d) Which is a globular cluster?

   ___ (e) Which has many bright blue stars?

   ___ (f) Which may contain up to 10 million stars?

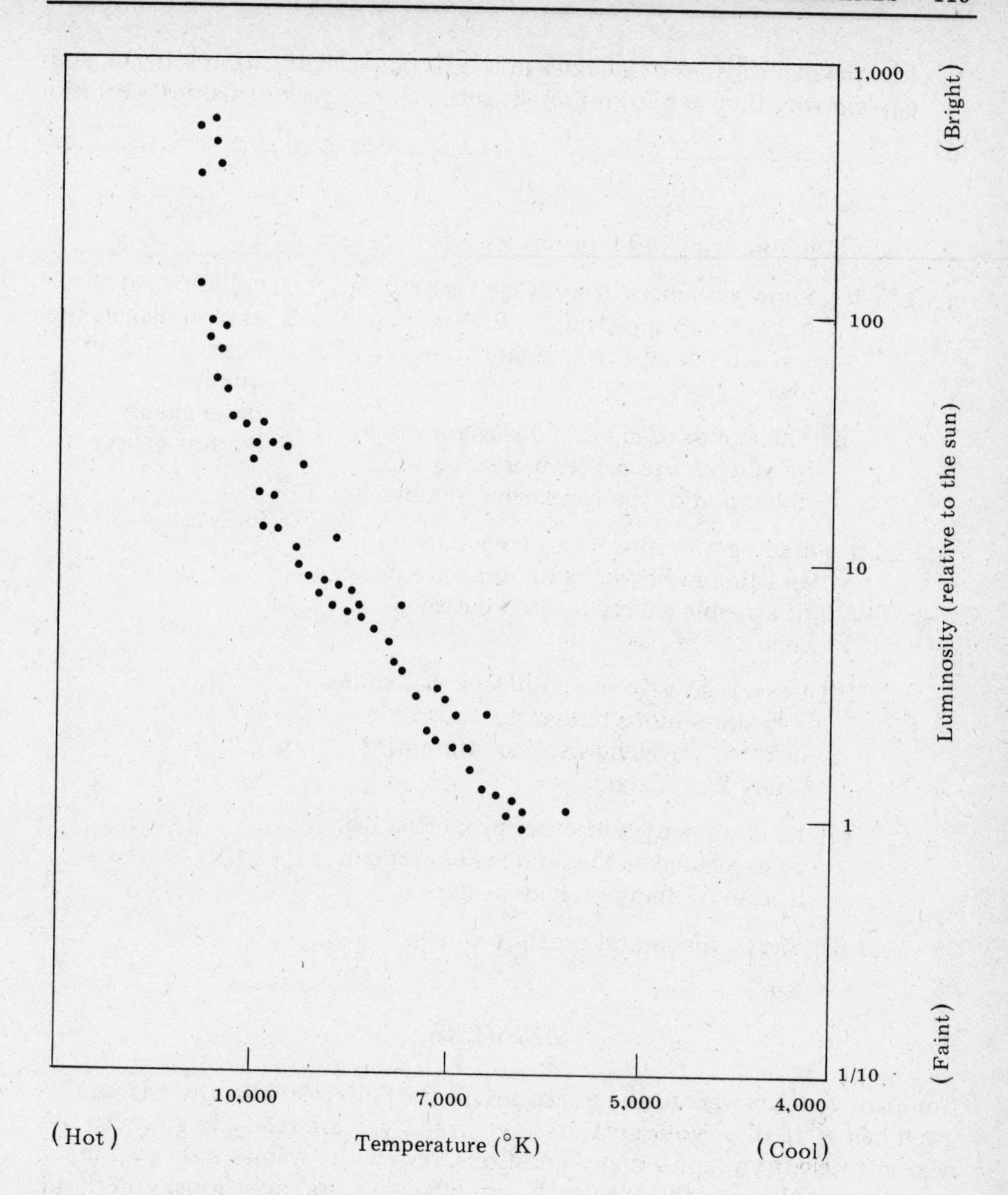

Figure 5.23. H-R diagram ("Cluster 2").

9. (a) What is the most distant object visible to the naked eye?

____

(b) How long does it take light emitted from that object to reach your eyes? ______________________________

10. List the main shapes of galaxies in the Hubble classification scheme, and explain why they cannot represent successive stages of galaxies' evolution.

____

____

11. Match the following descriptions with the correct object:

1. colliding galaxies
2. exploding galaxies
3. normal galaxy
4. quasar
5. radio galaxy
6. Seyfert galaxy

___ (a) Show streamers of hydrogen gas racing outwards at speeds up to 600 miles per second, perhaps from a violent event inside.

___ (b) The clouds of gas and dust here would be much more dense, increasing the likelihood of the formation of new stars.

___ (c) Radiograph shows two large patches emitting radio waves on opposite sides of a visible galaxy located between them.

___ (d) Has a relatively small nucleus that shines 100 times more brilliantly, mostly in infrared wavelengths, than our entire Milky Way Galaxy.

___ (e) Its luminosity and other properties can be explained as the composite of a collection of many individual stars.

___ (f) Shows the largest redshift known.

ANSWERS

Compare your answers to the questions on the Self-Test with the answers given below. If all of your answers are correct, you are ready to go on to the next chapter. If you missed any questions, review the frames indicated in parentheses following the answer. If you missed several questions, you should probably reread the entire chapter carefully.

1. an enormous collection of stars and gas and dust held together in space by the force of gravity (frame 1)
2. planet, star, solar system, open cluster, galaxy, cluster of galaxies, supercluster (frames 1, 2, 8, 16)

3.

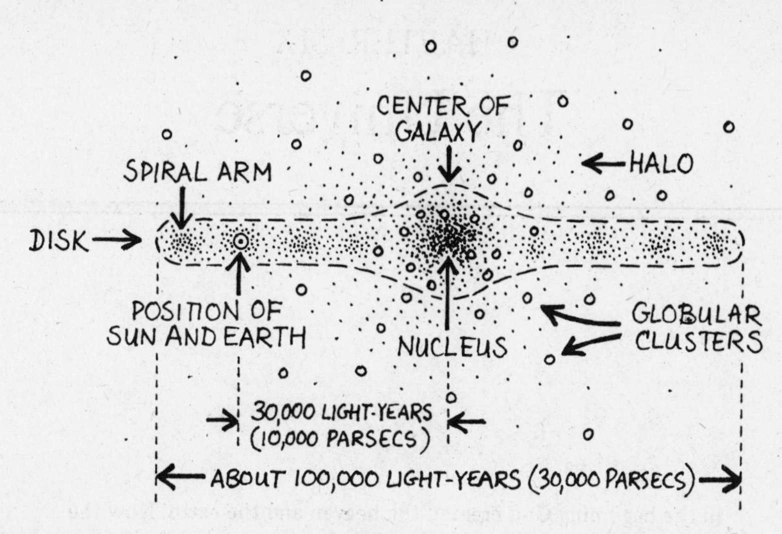

Figure 5.24. Edge-on view of the Milky Way.

(frames 2, 3, 11)

4. Hydrogen gas, radiation, tiny solid dust particles, water vapor, gases of elements heavier than hydrogen, organic molecules. (Frame 4)

5. Interstellar matter is the raw material for new stars and planets. (frames 4, 5)

6. No. The "horse's head" is a dark nebula. It is a relatively dense concentration of interstellar matter whose dust absorbs or scatters starlight and hides stars that are behind it from our view. (frame 6)

7. Radio waves pass through the interstellar dust in the disk of the Milky Way Galaxy much more effectively than visible light waves. (frame 7)

8. (a) 1; (b) 2; (c) 2; (d) 1; (e) 2; (f) 1 (frames 8, 9, 10)

9. (a) Andromeda Galaxy; (b) 2 million years (frame 12)

10. Elliptical, spiral, barred-spiral, irregular. All contain old stars, so all must be equally old. (frames 13, 14, 15)

11. (a) 2; (b) 1; (c) 5; (d) 6; (e) 3; (f) 4 (frames 3, 17, 18, 19, 20)

CHAPTER SIX

The Universe

> **In the beginning God created the heaven and the earth. Now the earth was unformed and void, and darkness was upon the face of the deep; and the spirit of God hovered over the face of the waters. And God said: "Let there be light." And there was light. And God saw the light, that it was good.**
>
> **Genesis 1:1-4**

1. People have always wondered about how the world began and if it will end. *Cosmology* is the branch of science concerned with the origin, present structure, evolution, and final destiny of the universe.

Astronomers construct hypotheses, called *cosmological models*, that try to explain how the universe began, how it is changing as time goes by, and what will happen to it in the future. These models must be consistent with the observational data we have on stars and galaxies.

Cosmological models differ from religious explanations of the universe in a very basic way. Can you state the difference? ______________________

____

\- - - - - - - - - - - - - - - - -

Cosmological models do not give a supernatural cause or meaning to physical events but try to explain these events using only the laws of nature.

2. The basic observation that must be accounted for by any cosmological model is that light from distant galaxies is shifted in wavelength toward the red end of the spectrum. This phenomenon is called the *cosmological redshift*

Modern theory says this redshift is a Doppler effect (frame 3.9), indicating that other galaxies are racing away from us. The most distant galaxies we observe have the greatest redshifts. They are moving away fastest of all.

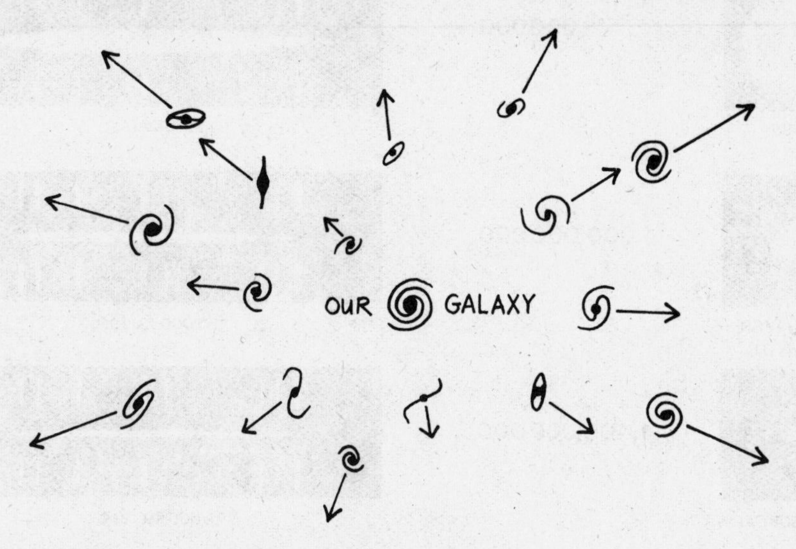

Figure 6.1. Galaxies receding from the Milky Way.

Whenever we look out into space we see galaxies receding from us. What does this observation implý about the universe? ____________________

____

- - - - - - - - - - - - - - - - - - - -

The universe must be *expanding.*

3. Examine figure 6.2, which shows the redshifts and corresponding calculated velocities for five galaxies that are at different distances away from us.

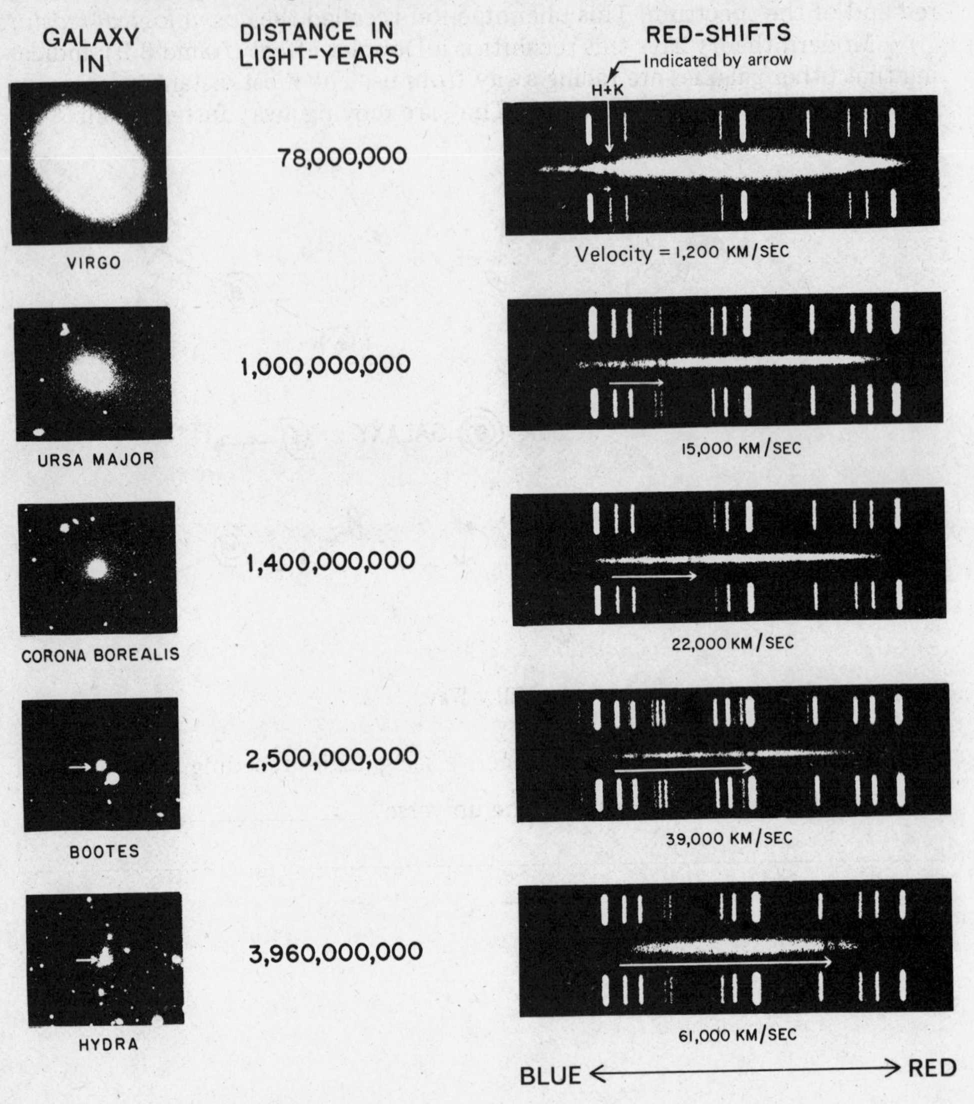

Figure 6.2. Redshifts and corresponding velocities of five galaxies.

Laboratory spectral lines of known wavelength are shown above and below the spectral lines of each galaxy for reference. A pair of absorption lines is marked at the top left of the reference spectrum in their unshifted positions. These two lines are shifted toward the red (to the right in the photograph) by increasing amounts for more distant galaxies.

Using figure 6.3a, plot a rough graph where each point represents the velocity of recession and distance of one of these galaxies. What do you ob-

80,000
70,000
60,000
50,000
40,000
30,000
20,000
10,000
0
Velocity (kilometers/second)
2 billion
4 billion
Distance (light-years)

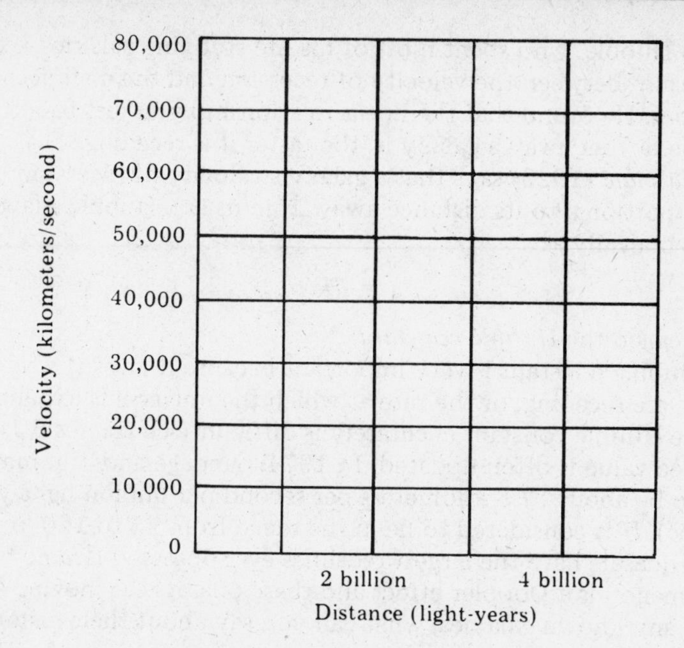

Figure 6.3a.

serve when you draw a smooth curve through the five points?__________

____

Explain.____

____

- - - - - - - - - - - - - - - - - - - -

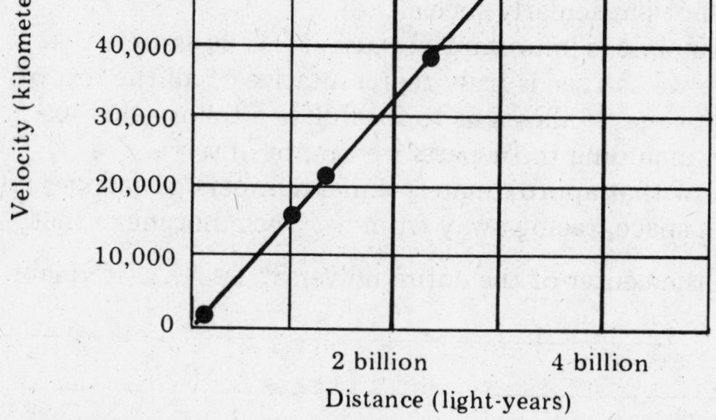

80,000
70,000
60,000
50,000
40,000
30,000
20,000
10,000
0
Velocity (kilometers/second)
2 billion
4 billion
Distance (light-years)

The points all lie near a straight line. That means there is a linear relationship between velocity of recession and distance away from us for these galaxies.

Figure 6.3b.

4. Edwin Hubble, who spent most of his life studying galaxies, examined the relationship between the velocity of recession and the distance away for many galaxies. He found that the linear relationship you just found is true in general: The farther away a galaxy is, the faster it is receding.

Hubble's law (1929) says that a galaxy's velocity of recession (v) is directly proportional to its distance away from us (x). Hubble's law can be written algebraically as

$$v = Hx$$

where H is called the *Hubble constant.*

The Hubble constant is very important because it gives the rate at which the galaxies are receding, or the rate at which the universe is expanding. To measure the Hubble constant accurately is difficult (see frames 6.10 and 6.11), and its stated value is often updated. In 1974, Sandage and Tammann reported the value to be about 17.5 kilometers per second per million light-years (57 km/sec/Mpc). H is considered to lie in the range from 45 to 120 km/sec/Mpc.

Some quasars have the largest redshifts ever observed (frame 5.20). If this phenomenon is a Doppler effect and these quasars are moving away faster than any known galaxies, what can you say about their distance?

____

Explain. _______________

____

- - - - - - - - - - - - - - - - -

These quasars are the most distant objects we can observe. Hubble's law says the most distant objects are those that are moving away the fastest.

5. The basic assumption we make in attempting to understand the cosmos is called the *cosmological principle.* The cosmological principle states that on the large scale the cosmos is the same everywhere at any given time.

According to this principle, any observer anywhere in the universe would see at any given time just about the same things we do on a large scale. Our neighborhood in space is not particularly special.

The cosmological principle is important because it lets us assume that the small portion of space we can see is truly representative of all the rest of the universe that we cannot see. It allows us to formulate a theory that explains the entire universe, including those parts we cannot observe.

Our observations show that approximately equal numbers of galaxies are located in all directions in space, racing away from us. Does that mean that our Milky Way Galaxy is the center of the entire universe? _______ Explain.

____

____

- - - - - - - - - - - - - - - - -

No. The cosmological principle says that if you went to any other galaxy and looked out into space, you would still see approximately equal numbers of galaxies located in all directions in space, racing away from you.

You can do a simple experiment to illustrate this principle. Obtain a balloon. Mark dots randomly on the balloon to represent galaxies. Label one dot *MW* to represent our Galaxy. Blow up the balloon. Observe how all the dots (galaxies) recede from each other.

6. One cosmological model, the *big bang theory*, says that the universe began with a big explosion over 18 billion years ago. All of the matter of our present universe was originally packed together in the *primeval fireball*—an extremely hot, dense ball that exploded with a tremendous big bang. The explosion threw hydrogen, helium, electrons, and radiation out into space.

The matter that was thrown out into space expanded and cooled. Several million years later, it condensed into galaxies. The universe has continued to expand, and the galaxies have continued moving away from each other ever since.

Today, as we have observed, the universe is still expanding. Stars are still forming inside galaxies, using the original hydrogen from the big bang.

In the future, the original hydrogen will finally be used up in stars. Then the stars and galaxies will all stop shining. The universe which began with a fiery big bang will fade into darkness with a cold "whimper" if it continues to expand indefinitely.

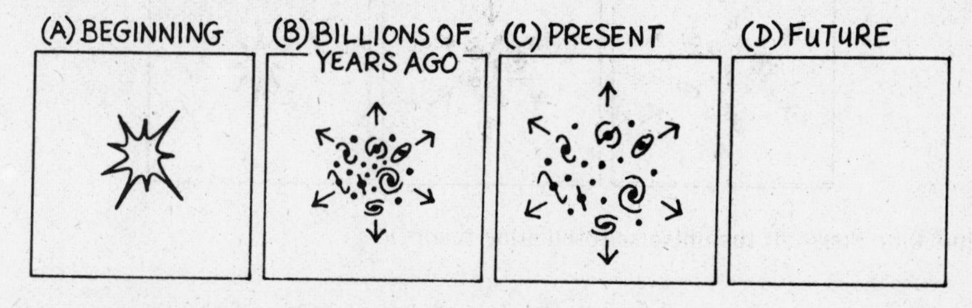

Figure 6.4. Stages of the universe (big bang theory).

Referring to figure 6.4, identify and briefly describe the stages of the universe according to the big bang theory.

(a) ____________________; (b) ____________________;

(c) ____________________; (d) ____________________

(a) Big bang explosion took place. (b) Galaxies formed out of ejected matter. (c) Galaxies are still receding; the universe is expanding. (d) Original hydrogen will be used up; a cold, black universe will result.

7. A second model, the *oscillating theory*, says that our universe began with a big bang but will not expand forever. Gravity will halt the expansion.

This model says that the universe was always oscillating, or expanding outward and contracting back inward, in the past and will always be oscillating in the future.

We happen today to be in the expanding universe phase, as has been observed. Our universe has been expanding for about the past 18 billion years, ever since the tremendous explosion of matter described by the big bang theory.

In the future our expanding universe will slow down, come to a complete stop, and then begin to contract. As it contracts, galaxies will fall back inward toward one another until all matter is once again packed very close together. Then another big bang will occur, and a new expanding universe will be born out of the same matter. The universe oscillates forever.

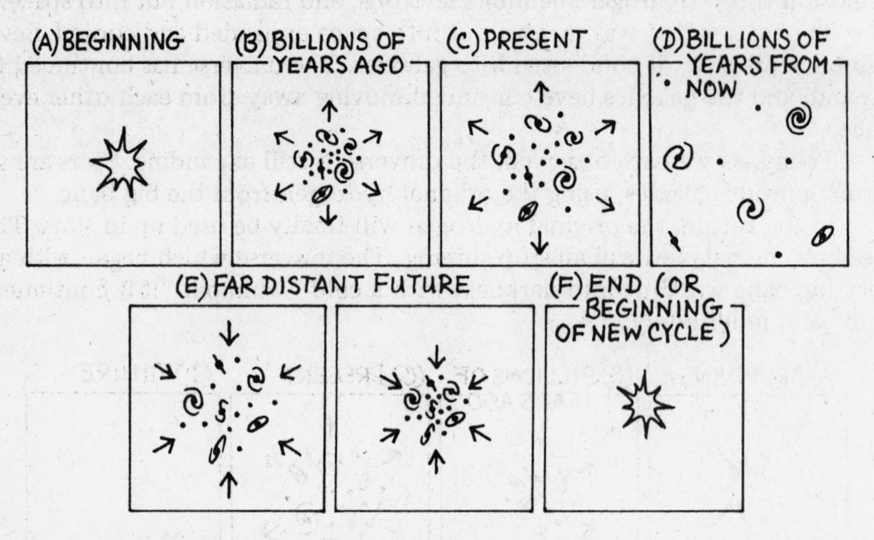

Figure 6.5. Stages of the universe (oscillating theory).

Referring to figure 6.5, briefly describe the stages of the universe according to the oscillating theory.

(a) ______________________________;

(b) ______________________________;

(c) ______________________________;

(d) ______________________________;

(e) ______________________________;

(f) ______________________________.

- -

(a) Big bang occurred. (b) Galaxies formed and continued to recede. (c) We live in an expanding universe; galaxies are racing away from one another. (d) Galaxies will stop. (e) The universe will contract; galaxies will fall back inward. (f) Matter will be packed together again. (g) A new big bang will occur; a new universe will be born.

8. A third model, the *steady state theory*, says that the universe does not evolve or change in time. There was no beginning in the past and there will be no end in the future. Past, present, future—the universe is the same forever.

This model assumes the *perfect cosmological principle.* This principle says that the universe is the same *everywhere* on the large scale, at *all times.* It maintains the same average density of matter forever.

In order to explain the observation that the universe is expanding, this model says that new hydrogen is created continuously in empty space at a rate just sufficient to replace matter carried away by receding galaxies.

Many astronomers dislike the steady state model because it says that new hydrogen is continuously created without explaining where the new hydrogen comes from. Such creation violates a basic law of physics—the *law of conservation of energy*—which states that the total energy in an isolated system always remains the same. Energy cannot be created or destroyed, although transformations may occur within the system.

Those astronomers who favor the steady state model like it because of its philosophical appeal. It defines a universe that always existed in the past and will always exist in the future.

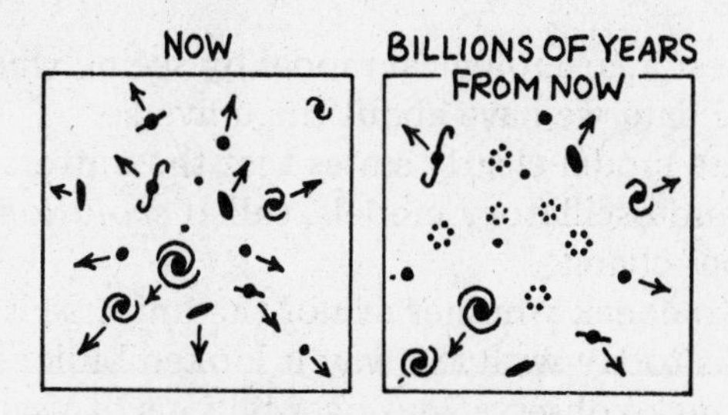

Figure 6.6. Stages of the universe (steady state theory).

Using figure 6.6, briefly describe the universe according to the steady state model.

(a) ______________________________

(b) ______________________________

- - - - - - - - - - - - - - - - - -

(a) Galaxies are receding, the universe is expanding, new matter is being created, new galaxies are being formed.

(b) The same pattern will occur. The universe maintains the same average density forever.

9. Write a short summary describing how each of the three cosmological models we have discussed answers the question: "Did the universe have a beginning in the past, and will it end in the future?" ____________

____

____

____

____

— — — — — — — — — — — — — — — —

Your answer should include the following ideas:

The big bang theory: The universe began with a tremendous explosion about 18 billion years ago and will become cold and dark in the future when the original hydrogen is all used up in the formation and evolution of stars.

The oscillating theory: Our present universe began with a big bang about 18 billion years ago, is still expanding now, and in the future will stop, contract, and end, but a new explosion will occur, and a new universe will be born out of the same matter.

The steady state theory: The universe had no beginning and will have no end but stays the same forever. New hydrogen is continuously created to replace matter carried away by receding galaxies, so the universe always stays the same.

10. Astronomers test a cosmological model by seeing whether it agrees with all the observational data we have about the universe.

The steady state model clearly states that the universe never changes, while the big bang and oscillatory models, called *evolutionary models*, say that it definitely does change.

The best way to check whether or not the universe is evolving is to compare the way it looks today with the way it looked billions of years ago. Since we cannot actually make observations over billions of years as the universe ages, astronomers instead look at galaxies that are at different distances away from us.

Although the idea—to look back in time you study current photographs of distant galaxies—is simple, it is very hard to carry out in practice. Today's technology is not sufficiently developed to allow detailed photography of very distant objects. Consequently, all data which might be used to check the cosmological models are full of uncertainties. No sufficiently precise data are yet available to confirm that any one of the models is definitely correct.

How can astronomers find out what the universe was like (a) 2 million years ago? ____________

(b) 3 billion years ago? ____________

Explain. ____

____

____

- - - - - - - - - - - - - - - - - -

They can examine photographs of galaxies such as (a) Andromeda, which is 2 million light-years away from Earth, and (b) Hydra, which is about 3 billion light-years away. It takes light one year to travel about 6 trillion miles, or one light-year. The light we now receive on photographic plates left Andromeda 2 million or Hydra 3 billion years ago; it tells us now what the universe was like then.

11. Astronomers use this method to compare the value of the Hubble constant now and billions of years ago.

The steady state theory says that the universe has always been expanding at the same rate it is today. The big bang theory says that the universe is slowing down now, long after the original explosion. The oscillating theory says that the universe is slowing down at such a rate that it will actually come to a complete stop.

If the galaxies are slowing down, as the evolutionary models say, then the Hubble constant will be smaller now than it was billions of years ago. The Hubble constant decreases faster in the oscillating theory than in the big bang theory. It does not change at all in the steady state theory.

Allan Sandage at Mount Palomar has been working on this problem for several years. He reports that the Hubble constant is decreasing, meaning that the universe is slowing down in its expansion. No one can measure distant galaxies with high precision yet, so his results are too uncertain to be absolutely conclusive.

Why is it so important to measure the value of the Hubble constant very accurately? ____

____

- - - - - - - - - - - - - - - - - -

An accurate value of the Hubble constant would be strong evidence in favor of one and against the other cosmological models described.

12. The evolutionary models predict that the universe should still be filled with *cosmic background radiation*—a small remnant of radiation left over from the original big bang. The primeval fireball would have sent strong short wave radiation (corresponding to a temperature of about 10 million °K) in all directions into space like an exploding atomic bomb. In time, that radiation would spread out, cool, and fill the expanding universe uniformly. By now it would strike Earth as microwave (short radio) radiation, corresponding to a temperature only a few degrees above absolute zero.

In 1965, microwave radiation like that emitted by a source whose temperature is about 2.7° K was detected coming equally from all directions in the sky. And so it appears that astronomers have detected the fireball radiation that was produced by the big bang.

What does the discovery of the fireball radiation mean for the steady state theory? ______________________________

- - - - - - - - - - - - - - - - - - -

It casts serious doubt on the steady state model. The steady state model must find a way to explain the existence of this radiation, or else the model cannot be correct.

13. Observations of the density of matter in the universe and of quasars give other important clues for deciding among the alternative cosmological models.

The *density of matter* in the universe required for the force of gravity to stop the expansion of the universe according to the oscillating model is 5×10^{-30} grams per cubic centimeter. Today the observed density of galactic material throughout the universe is not half of that required to stop the expansion. Some astronomers have been searching for what is called the *missing mass* for several decades. Unobserved dark matter could be neutrinos, black holes, unlit stars, or some other type.

If the quasars are really as far away as their redshifts indicate, many more of them existed in the past than do now. If events occurred when the universe was young that happen very rarely now, the perfect cosmological principle and steady state model cannot apply.

What would be the cosmological significance of new discoveries of so-far-unobserved mass and energy in the universe? ______________________________

____

- - - - - - - - - - - - - - - - - - -

The density of matter in the universe would be greater than the present calculated value. It might be sufficient to stop the expansion of the universe, as predicted by the oscillating model.

14. Estimates of the *age of the universe* have tended to grow from a biblical few thousand years to millions and then billions of years. A recent estimate by David N. Schramm at the University of Chicago put the age of the universe at 20 billion years. This estimate was based on the method of radioactive dating using rhenium 187. It involved calculating how much radioactive rhenium has decayed since that element was first formed early in the history of the Milky Way Galaxy.

Other estimates of the age of the universe are based on the value of the Hubble constant. The *Hubble time*, which is the age of the universe since the time of the big bang, is equal to $\frac{1}{H}$. The calculated age depends greatly on the value of the Hubble constant, which is still not precise, with some correction for the slowing down of the universe in the past. This method puts the time of the big bang at from 10 to 20 billion years ago.

Still another age is derived from that of the oldest stars. By this method, the universe is from 9 to 18 billion years old.

List three methods of estimating the age of the universe. ___________

____

____

____

– – – – – – – – – – – – – – – – – – –

1. radioactive (rhenium 187) dating;
2. measuring the Hubble constant and Hubble time of $\frac{1}{H}$;
3. estimating based on the age of the oldest stars observed

15. Estimates of the radius of the universe also depend very much on the value of the Hubble constant. The *Hubble distance*, the distance to the edge of the observable universe, is equal to the speed of light divided by the Hubble constant, $\frac{c}{H}$. These estimates put the radius of the universe between 12 billion and 16 billion light-years.

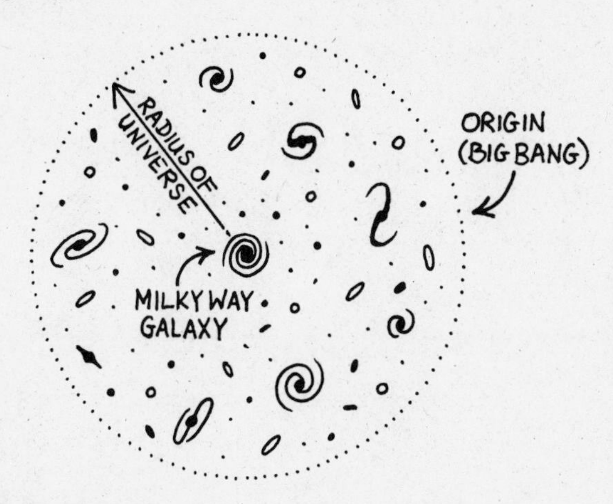

Figure 6.7. The Hubble distance.

The eternal questions of mankind—how did the world begin, and will it end?—cannot be answered by science today. Review the range of estimates we have about the universe today by filling in the following chart:

The Observable Universe Today

(a) Rate of recession of distant galaxies (Hubble constant)

____

(b) Approximate radius

____

(c) Age in approximately present form

____

(a) some 17.5 kilometers per second per million light-years, or 57 km/sec/Mpc (a rough value). The rate of recession is considered to lie in the range 45 to 120 kilometers per second per million parsecs.

(b) from 12 billion to 16 billion light-years

(c) from about 10 billion to 20 billion years old

SELF-TEST

This self-test is designed to show you whether or not you have mastered the material in Chapter Six. Answer each question to the best of your ability. Correct answers and review instructions are given at the end of the test.

1. Define cosmology. ____________________

____

2. How do cosmological models differ from religious explanations of the universe? ____________________

3. Describe the evidence that the universe is expanding. ____________________

____

4. Match one *or more* of the three main cosmological models with each statement:

    1. big bang theory
    2. oscillating theory
    3. steady state theory

    ____ (a) Roughly 18 billion years ago the universe exploded from a dense hot ball of matter.

    ____ (b) Matter is being continually created in space between the galaxies.

    ____ (c) Galaxies are moving away from each other with speeds that increase with distance.

    ____ (d) The universe is the same everywhere, on the large scale, at all times.

    ____ (e) The universe goes through an endless series of expansions and contractions.

    ____ (f) The rate of expansion of the universe is decreasing at the present time.

5. State Hubble's law. ____________________

____

6. Why is the Hubble constant so important in cosmology? ____________________

____

7. List at least two basic observations which can help decide among the big bang, the steady state, and the oscillating theories. ____________________

____

8. What is the approximate (a) Hubble age of the universe?

____

(b) Hubble radius? ______________________________

9. Which type of cosmological model, evolutionary or steady state, is most favored today based on observed data? ______________________

ANSWERS

Compare your answers to the questions on the self-test with the answers given below. If all of your answers are correct, you are ready to go on to the next chapter. If you missed any questions, review the frames indicated in parentheses following the answer. If you miss several questions, you should probably reread the entire chapter carefully.

1. Cosmology is the branch of science concerned with the origin, the present structure, evolution, and final destiny of the universe. (frame 1)

2. Cosmological models do not give a supernatural cause or meaning to physical events, but try to explain these events using only the laws of nature. (frame 1)

3. Light from distant galaxies is shifted in wavelength toward the red end of the spectrum, a phenomenon called the redshift. The farther away a galaxy is, the greater its redshift. The most distant galaxies we observe have the greatest redshifts of all, indicating that the galaxies are receding uniformly from us and from each other. (frame 2)

4. (a) 1 and 2; (b) 3; (c) 1, 2, and 3; (d) 3; (e) 2; (f) 1 and 2
(frames 2, 5, 6, 7, 8, 11)

5. Hubble's law (1929) says that a galaxy's velocity of recession (v) is directly proportional to its distance away from us (x). Hubble's law can be written algebraically as $v = Hx$, where H is called the Hubble constant. (frame 4)

6. The Hubble constant is very important because it gives the rate at which the galaxies are receding, or the rate at which the universe is expanding. It is a basis for estimating the size and age of the universe.
(frames 4, 14, 15)

7. Any two of the following:
   (a) rate of change of the Hubble constant with time
   (b) fireball radiation
   (c) density of matter in the universe
   (d) distance to quasars

   (frames 11, 12, 13)

8. (a) 10 billion to 20 billion years
   (b) 12 billion to 16 billion light-years

   (frames 14, 15)

9. evolutionary (frames 11, 12, 13)

CHAPTER SEVEN
The Sun

Thou dawnest beautifully in the horizon of
the sky,
O living Aton who wast the Beginning of life!

Ikhnaton (c. 1385-1358 B.C.),
Hymn to the Sun

1. The sun is the star closest to Earth. It provides the light, heat, and energy for life.

Ancient peoples worshipped the sun as a life-giving god. Some of the names given to the sun god were Aton, Helios, and Sol. Modern astronomers study the sun today. It is critical to Earth and is a key to understanding distant stars that cannot be observed in detail.

The *sun's total energy output* is enormous (the sun's luminosity is 3.83×10^{26} watts). Solar energy is practically inexhaustible. The amount of the sun's energy that falls per second on Earth's outer atmosphere, called the *solar constant*, is about 126 watts per square foot. This amount of energy provides about as much heat and light in a week as is available from all of our known reserves of oil, coal, and natural gas.

Our sun is dynamic and seething. It is in turn extraordinarily active and relatively quiet. Changes in solar energy output affect Earth's climate, atmosphere, and weather, as well as modern power-transmission and communication systems. These changes are monitored to gain a better understanding of the sun's effect on Earth.

State three reasons why modern astronomers, physicists, and engineers are using their most sophisticated techniques to determine the true nature of the sun. ____

____

- - - - - - - - - - - - - - - - - -

(1) The sun is an almost inexhaustible source of present and potential future energy. (It is also free and nonpolluting.)

(2) The sun is the only star close enough to observe in detail, so astronomers use it to determine what other stars are like.

(3) Changes in the sun's energy output affect Earth's climate, atmosphere, and weather, as well as power-transmission and communication systems.

2. The average *distance* between the Earth and the sun, called the *astronomical unit* (AU), is about 93 million miles (officially 149,597,870 km). Astronomers use this astronomical unit as a measure of distance in the solar system.

The sun is a huge gaseous sphere with a radius of about 432,000 miles (696,000 km). It looks about the same size as the full moon in the sky, with an angular diameter of 32′ (about $\frac{1}{2}°$), because of its much greater distance from Earth.

Warning: You should *never* look directly at the sun through binoculars or a telescope. You could permanently injure your eyes. The safest way to

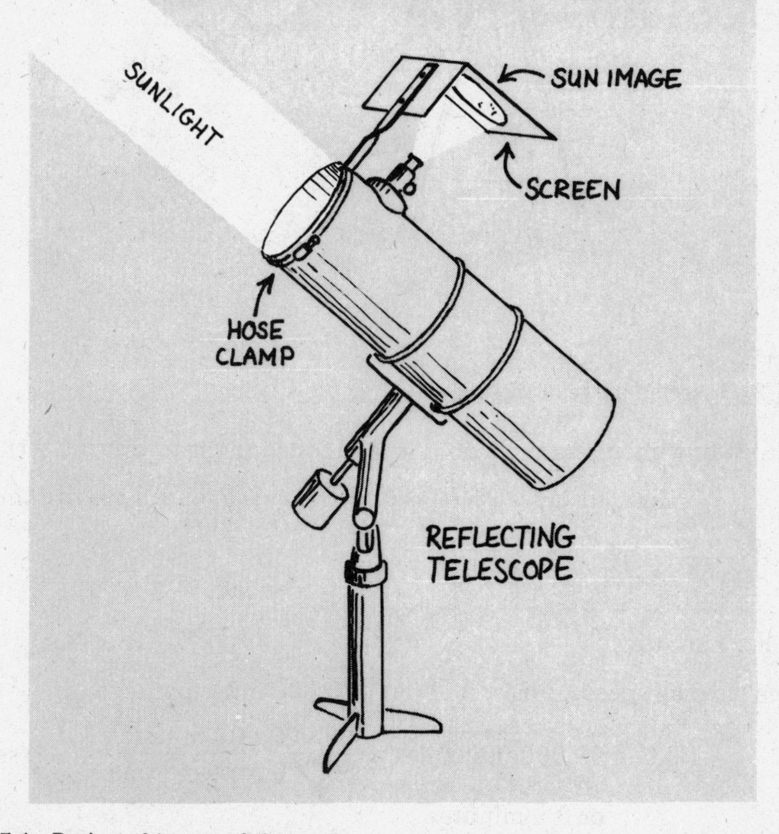

Figure 7.1. Projected image of the sun.

observe the sun is to project its image onto a white card or paper held behind the eyepiece. Finer details may be observed using a high-quality, special solar filter (available to amateurs) that transmits only enough light to form a clear image of the sun.

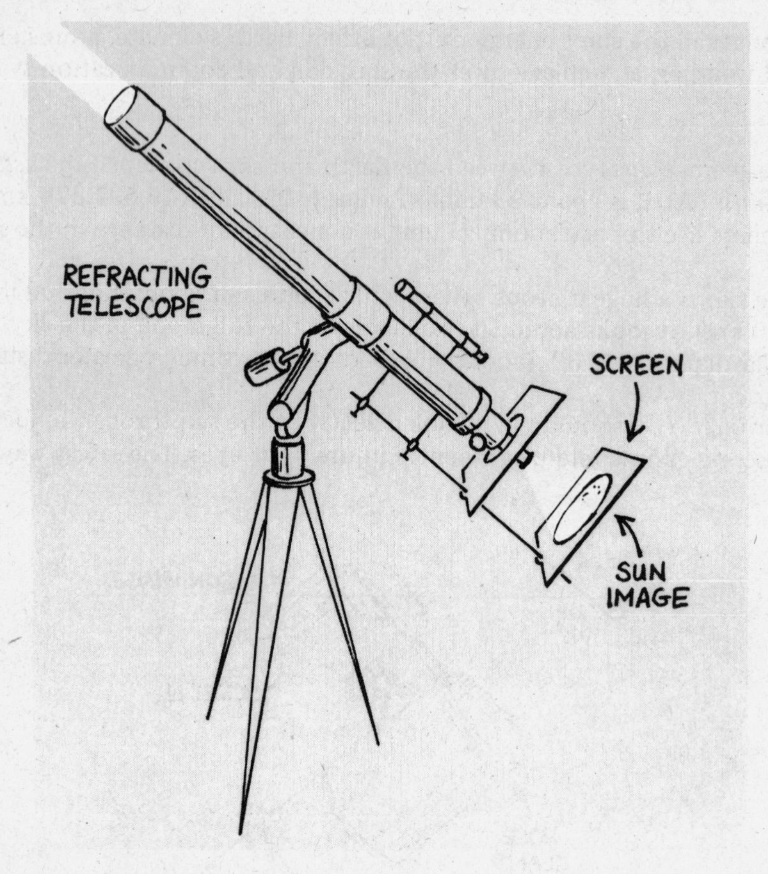

Figure 7.2. Projected image of the sun.

About how many minutes does it take for sunlight to travel 1 AU?

__________ Hint: distance = (speed) x (time). You can rewrite this:

$$\text{time} = \frac{\text{distance}}{\text{speed}}.$$

About $8\frac{1}{3}$ minutes.

Solution: The speed of light is about 186,000 miles per second.

$$1 \text{ AU} = 93{,}000{,}000 \text{ miles.} \quad \frac{93{,}000{,}000 \text{ miles}}{186{,}000 \text{ miles/second}} = 500 \text{ seconds,}$$

or $8\frac{1}{3}$ minutes.

(That means that if the sun stopped shining, you would not know about it until $8\frac{1}{3}$ minutes later.)

3. You learned about the *life cycle* of stars like our sun in Chapter Four (review frame 4.13, if necessary). The *condensation theory* of the origin of the solar system says that our sun and its family of planets were born together in a whirling cloud of gas and dust about 5 billion years ago.

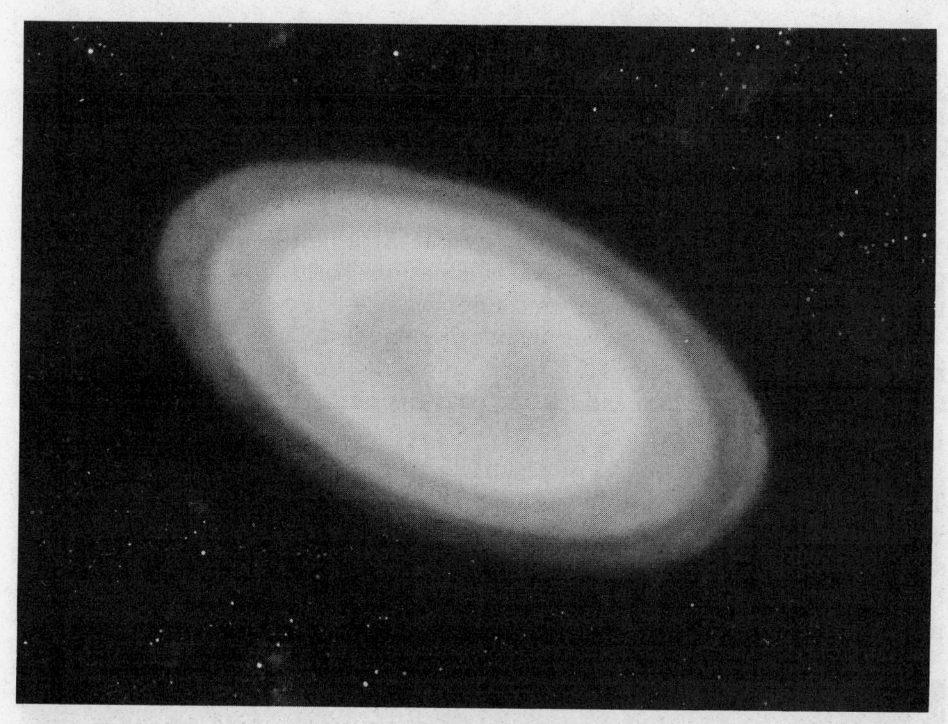

Figure 7.3. Artist's conception of condensing nebula.

Most of that cloud condensed into the sun. The sun has more than 99 percent of the *mass* of the solar system, $M_\odot = 1.989 \times 10^{33}$ grams, and provides the gravitational force that keeps the planets circling it.

More than sixty *chemical elements* have been identified in the sun's spectrum. The sun's outer layers are believed to have the same chemical composition as the original sun, about 71 percent hydrogen, 27 percent helium and 2 percent other elements by weight. The sun's core is believed to have been changed to about 38 percent helium in nuclear fusion reactions during the past 4.5 billion years.

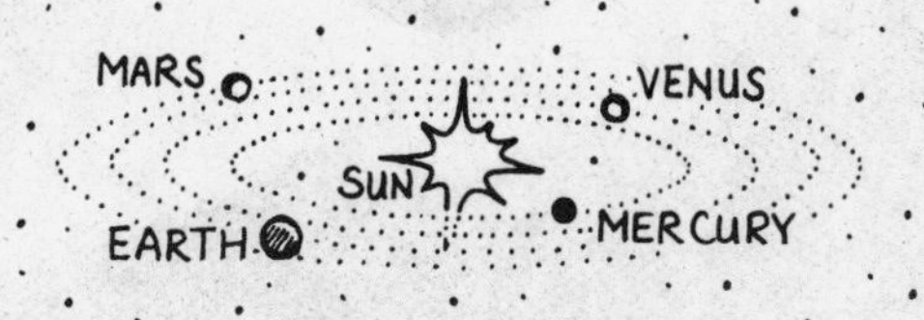

Figure 7.4. Possibly a condensing nebula gave birth to the planetary system surrounding the sun.

Why do astronomers expect to find other stars that have planets circling around them? ______________________________

____

The condensation theory says that the planets circling the sun were born together with their star. Since the sun is a typical star, it seems likely that other, similar stars were also born together with a family of planets.

4. Our picture of the sun's *structure* comes from direct observations of its outer layers plus indirect theoretical calculations based on the behavior of gases deep in its interior which we cannot see.

The sun's *core* is the power plant where nuclear fusion reactions generate the sun's energy (frame 4.5). Right below the surface is a region called the *zone of convection*, where circulating currents of gas transfer heat from the hot interior to the surface.

The *photosphere* is the visible surface of the sun that we see in the sky. The photosphere is a hot, thin gas layer about 5,800° K (10,000° F) from which energy is radiated into space. The *limb* is the apparent edge of the sun's disk. It looks darker than the center, an effect called *limb darkening*, because light from the limb comes from higher, cooler regions of the photosphere.

Figure 7.5. Total eclipse of the sun, during which the corona becomes strikingly visible.

The *chromosphere* is the lower part of the sun's atmosphere. It extends a few thousand miles above the photosphere. It is normally visible from Earth only during a total eclipse of the sun, when it glows red due to its hydrogen gas. The chromosphere contains other elements, including helium and calcium.

The *corona* is the sun's outer atmosphere just above the chromosphere. It is a thin, hot (1–2 million°K) gas that extends for millions of miles from the sun. The corona becomes strikingly visible during a total eclipse of the sun when the much brighter photosphere is briefly blocked from view.

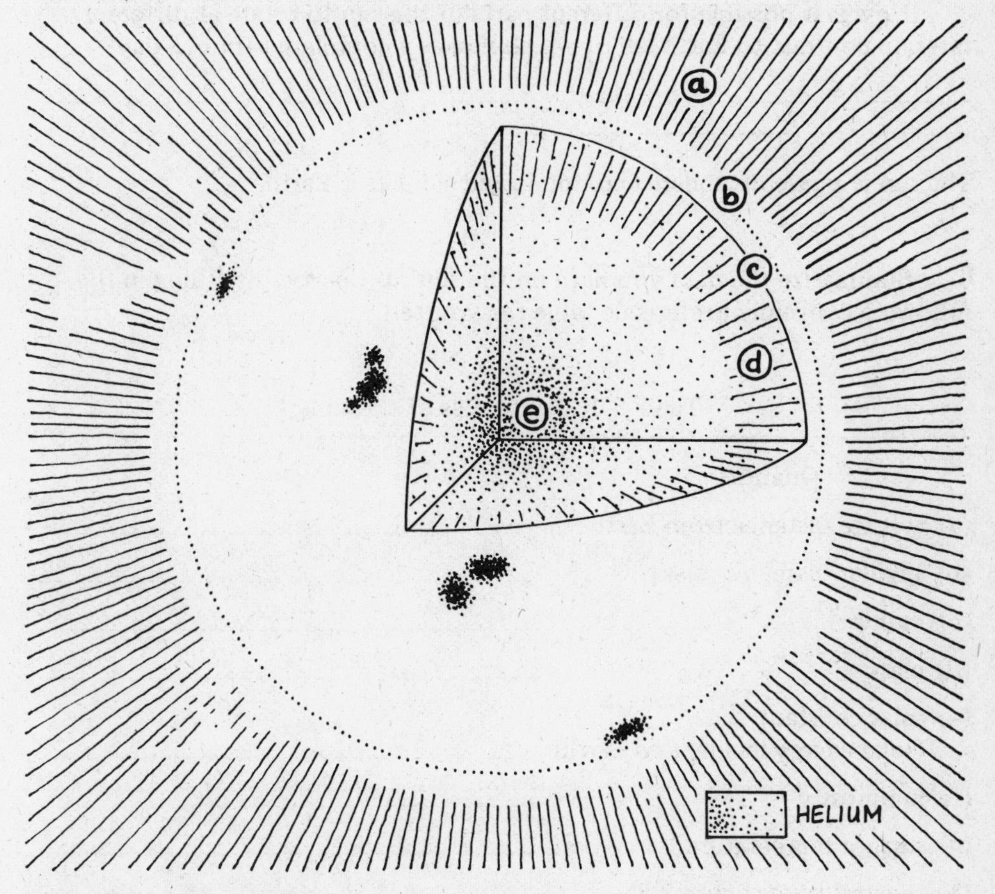

Figure 7.6. Regions of the sun.

Identify the regions of the sun lettered on figure 7.6.

(a) ____________; (b) ____________; (c) ____________;

(d) ____________; (e) ____________

– – – – – – – – – – – – – – – – – –

(a) corona; (b) chromosphere; (c) photosphere; (d) zone of convection; (e) core

5. The sun keeps turning around on its axis in space, from west to east, as Earth does. But there is a difference. All of Earth makes a complete turn in a day. But the whole sun does not turn around together at the same rate.

The *period of rotation*, or length of time for one complete turn, is fastest at the sun's equator (about 25 days), slower at middle latitudes, and slowest at the sun's poles (about 35 days). This strange rotation pattern probably contributes to the violent activity that takes place on the sun, described in the frames which follow.

How is it possible for different parts of the sun to rotate at different rates, in contrast to Earth, all of which makes a complete turn in a day?

____

— — — — — — — — — — — — — — — — — —

The sun is a gaseous sphere and not a rigid solid as is Earth.

6. Summarize the data you have on the sun's properties by filling in the following convenient reference table for yourself.

Table 7.1. Properties of the sun

Quantity	Value
(a) average distance from Earth	__________
(b) angular diameter in sky	__________
(c) radius	__________
(d) mass	__________
(e) solar constant (solar energy incident on Earth)	__________
(f) luminosity	__________
(g) surface temperature	__________
(h) spectral type (figure 3.8)	__________
(i) apparent magnitude (Table 3.3)	__________
(j) rotation period	__________
(k) chemical composition of outer layers	__________

— — — — — — — — — — — — — — — — — —

(a) about 93 million miles (149,597, 870 km); (b) 32′; (c) 432,000 miles (696,000 km); (d) 1.989×10^{33} grams; (e) 126 watts per square foot; (f) 3.83×10^{26} watts; (g) about 5,800° K (10,000° F); (h) G2; (i) −26.7; (j) equator: about 25 days; poles: about 35 days; (k) outer layers: about 71 percent hydrogen, 27 percent helium, 2 percent more than sixty other elements

7. Astronomers are using their most sophisticated tools and techniques to observe the sun more closely and in more detail than ever before. They have accumulated more data about the sun recently than in all of recorded history.

On the ground, special *optical solar telescopes* photograph the sun's visible surface with its changing features. Arrays of giant *radio telescopes* receive and record radio waves from different parts of the radio sun. *Infrared telescopes* observe the solar limb and map sunspots.

Figure 7.7. The 1.5 meter R.R. McMath solar telescope at Kitt Peak. A mirror at the top reflects sunlight down the long sloping tube.

Above Earth's atmosphere, *ultraviolet*, *X-ray*, and *gamma ray telescopes* on spacecraft record images of processes in the hottest and most active regions of the sun.

Formerly, the sun's chromosphere and corona could be observed directly only during the brief minutes of a total eclipse of the sun when the much brighter photosphere was blotted out (frame 4). Now astronomers do not have to wait for one of these rare natural events to occur. Color filters and *spectroheliographs* image the sun in light of essentially a single wavelength. *Spectroheliograms* are photographs of the sun in single color light belonging to one gas such as hydrogen or calcium. They reveal the distribution of different gases and local phenomena.

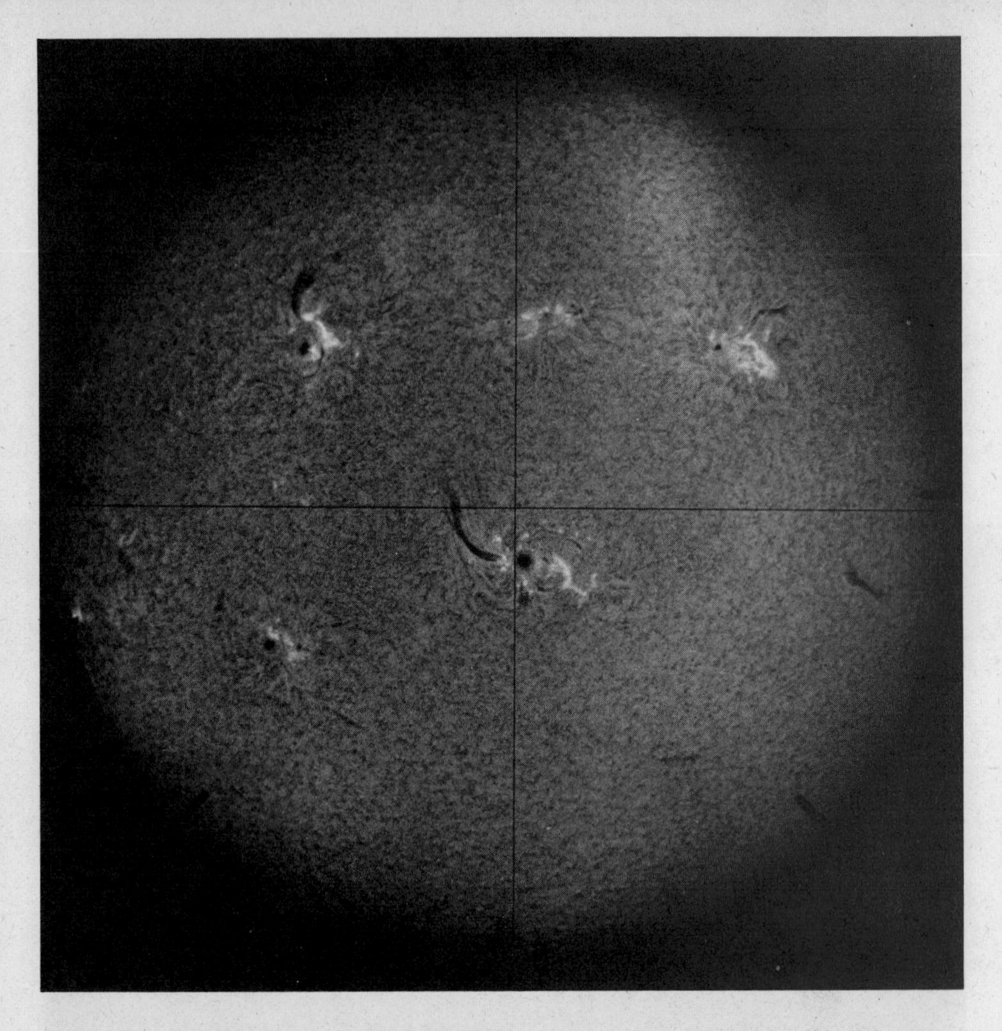

Figure 7.8. The sun photographed from Skylab in light of a spectral line of hydrogen (centered at 6563 Å).

Coronagraphs, telescopes designed to create an artificial eclipse, are used to photograph the corona. In 1973–1974, astronauts had a battery of eight solar telescopes aboard the *Skylab* space station 270 miles above Earth, and they observed the sun extensively. The Skylab coronagraph allowed $8\frac{1}{2}$ months of corona observation compared to less than 80 hours from all natural eclipses since use of photography began in 1839.

The most ambitious look at violent solar eruptions so far was the international *Solar Maximum Year* 1980–81 project. Hundreds of scientists, engineers, and solar observers in 18 countries using the newest telescopes, remote-sensing instruments, and spacecraft scrutinized the sun. Important data gathered in visible, ultraviolet, X- and gamma ray wavelengths promise new insights into our sun's complex nature.

Why do different features of the sun appear in pictures taken in light of different wavelengths such as visible light, ultraviolet rays, or X rays? Hint:

Review frame 2.10 if necessary. ______________________________

____

- - - - - - - - - - - - - - - - - - - -

Different wavelengths are produced in regions of different temperatures where different conditions and activities prevail.

8. An optical telescope reveals that the photosphere has a grainy appearance, called *granulation.* Bright spots that look like rice grains, called *granules,* usually dot about one-third of the sun's disk.

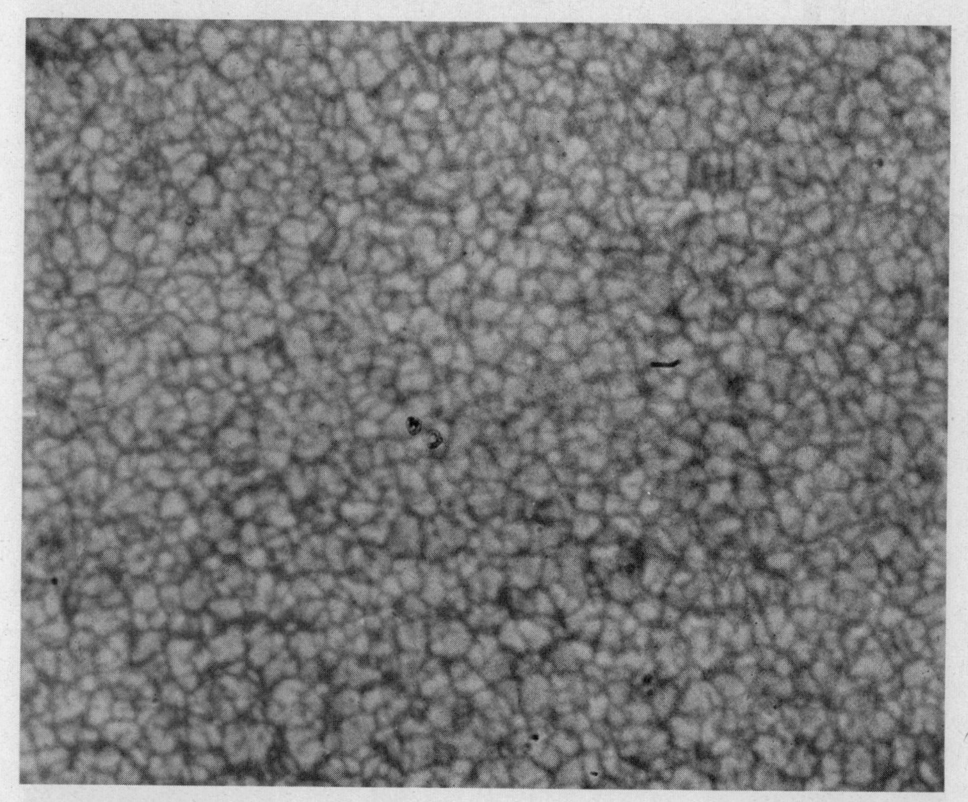

Figure 7.9. Solar granules.

Granules, about 625 miles (1,000 km) across, are about 300° hotter than the surrounding dark areas. Individual granules last an average of 5 minutes each. They are the tops of rising currents of gases from the sun's interior, seen through the photosphere.

Granules belong to *supergranules,* cells about 9,000-19,000 miles (15,000-30,000 km) across on the sun's disk. Supergranules last as long as a day. They have a flow of gases from their centers to their edges, in addition to the vertical gas currents in the granules.

Bright white patches called *faculae*, from the Latin for "little torches," may be visible near the sun's limb.

What causes granulation? ____

— — — — — — — — — — — — — — — — — —

gases rising from the sun's hot interior

9. *Sunspots* are temporary, dark, relatively cool blotches on the sun's bright photosphere. They usually appear in groups of two or more. Individual sunspots last anywhere from a few hours to a few months.

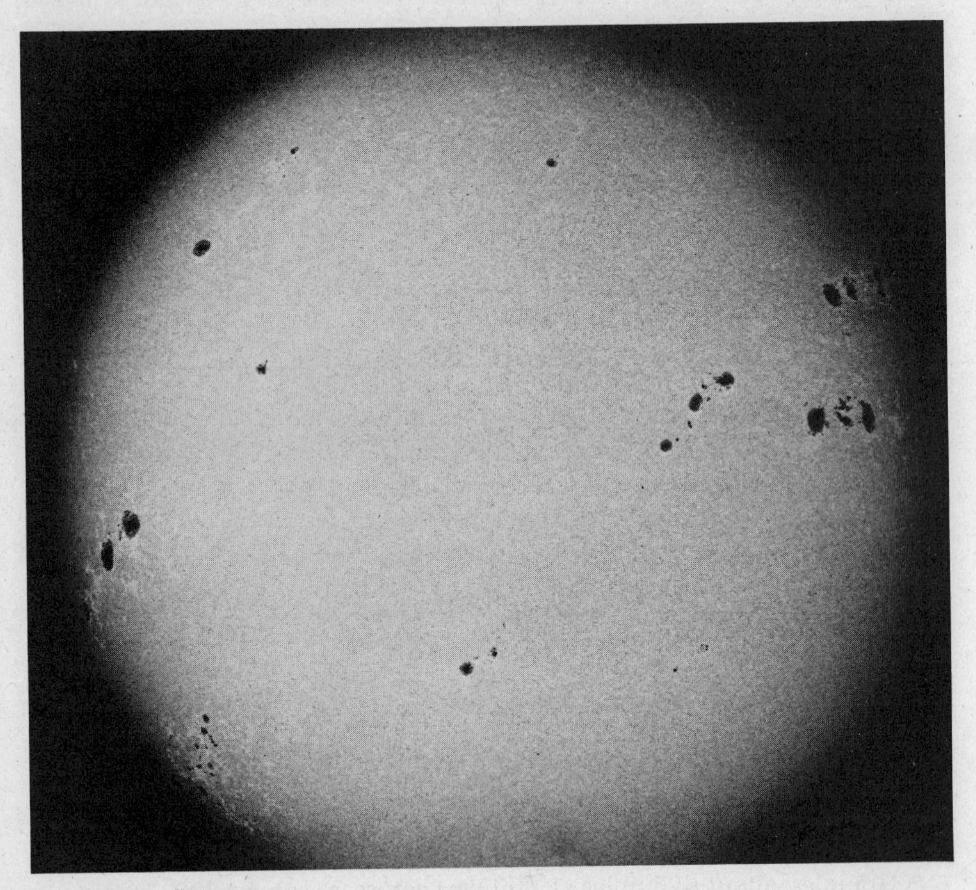

Figure 7.10. Sunspots.

The largest sunspots are visible at sunrise or sunset or through a haze. Observations of sunspots were first recorded in China before 800 B.C.

A typical sunspot is roughly as big as Earth. The largest sunspots that appear may be bigger than ten Earths.

Sunspots are really brighter and hotter than many stars. The temperature is about 4,200° K in the *umbra*, or core, and 5,700° K in the *penumbra*, or outer gray band of a large spot. They look dark only in comparison to the hotter, brighter surrounding photosphere.

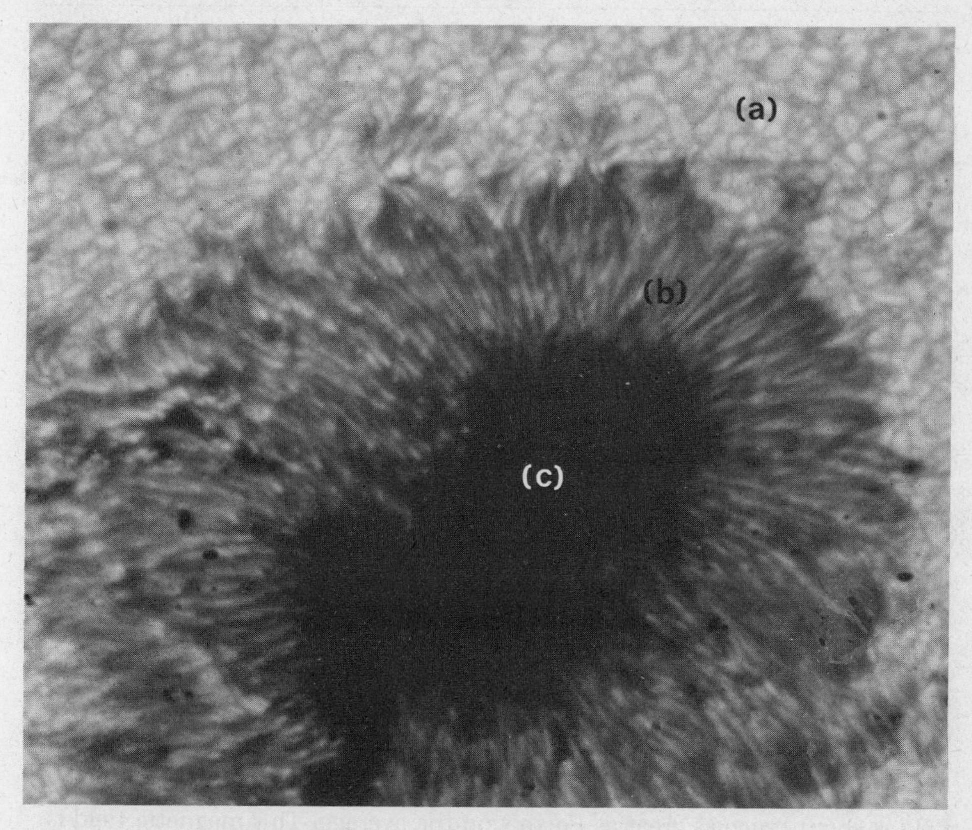

Figure 7.11. Umbra, penumbra of a sunspot; photosphere.

Identify the umbra, penumbra, and photosphere, lettered on figure 7.11, and indicate the approximate temperature of each.

(a) ________________; (b) ________________; (c) ________________

- - - - - - - - - - - - - - - - - - -

(a) photosphere, 6,000° K; (b) penumbra, 5,700° K; (c) umbra, 4,200° K

10. At any one time more than one hundred sunspots may appear on the sun's disk, or none at all. The number of sunspots repeatedly increases and decreases in a roughly regular cycle. The *sunspot cycle* is the approximately 11-year cycle of change in the number of visible sunspots.

The sunspot cycle is watched carefully from Earth as an indicator of solar activity. The sun is most active with greatest outbursts of energy and radiation during the years when sunspots are most numerous. It is least active in the years of sunspot minima. The 1980 Solar Maximum Year was the most widely scrutinized since Galileo made the first telescopic sunspot observation (frame 7.7).

Why is it important to keep track of the sunspot cycle? ______

______

The sun is most active during the years of sunspot maxima, pouring the greatest amount of energy and radiation into Earth's environment.

11. Sunspots are like huge *magnets.* They are the sites of powerful *magnetic fields,* or regions of magnetic force, that are typically more than one thousand times stronger than Earth's magnetic field.

The magnetic field of a sunspot can be detected before the spot itself can be seen and after the spot is gone. Therefore, magnetic fields probably shape and control local conditions on the sun.

The magnetic fields are created by motions of electrically charged particles that make up the sun's hot gases. Astronomers believe that most violent outbursts of material and radiation on the sun get their energies from these magnetic fields.

A weaker magnetic field spreads out over the whole sun. It has a north pole and a south pole, with the magnetic axis tilted 15 degrees to the rotation axis. It is split into two hemispheres. The sun's magnetic field probably extends from its northern hemisphere through the solar system out to Pluto (about 4 billion miles). Near the edge of the solar system, it bends and returns to the sun's southern hemisphere.

Over most of the sun's surface (except in active regions), its magnetic field is about twice as great as Earth's on the average. This magnetic field is generated by electrically charged particles inside the sun.

The sun's magnetic field has been observed to reverse its north and south poles about every 11 years near the time of sunspot maximum. It takes two sunspot cycles of about 11 years each for the magnetic poles to repeat themselves. So the sunspot cycle is sometimes counted as 22 years, rather than 11, when counting the length of time for one solar activity cycle.

What probably activates the violent outbursts of material that occur on the sun? ______

______

----

very strong magnetic fields at the sites of sunspots

You can observe a magnetic field by putting a magnet under a piece of paper. Lightly sprinkle iron filings on top of the paper. The filings will line up according to the strength of the magnetic force. By showing the regions of the magnetic force, they make the magnetic field visible to you.

12. The bright areas of solar activity that appear on spectroheliograms (figure 7.8) are called *plages*, from the French word for "beach." A plage surrounds a sunspot. It occurs in the chromosphere.

Prominences are flamelike masses of bright gas that rise hundreds of thousands of miles above the limb of the sun. They consist of relatively cool (10,000° K) dense gas in the hotter (10,000,000° K) thinner surrounding corona. Prominences seem to originate near sunspots.

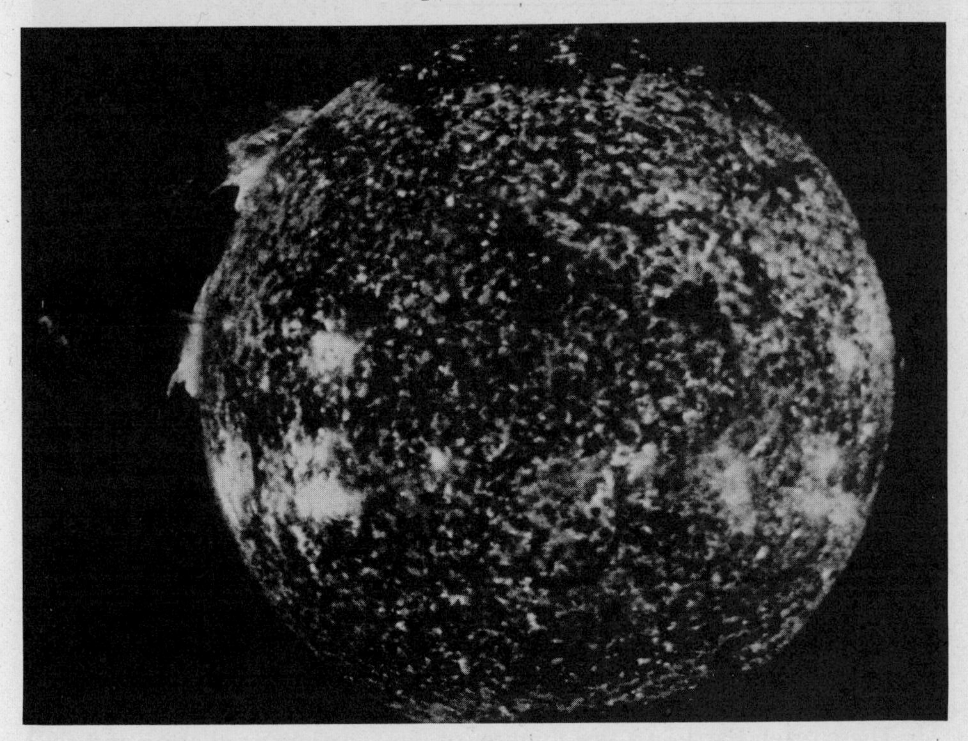

Figure 7.12. Several flares and a large prominence.

What probably bends and controls the gas in prominences? ____________

____

- - - - - - - - - - - - - - - - - - - -

strong magnetic fields in the vicinity of sunspots (see frame 7.11)

13. A solar *flare* is a sudden, tremendous, explosive outburst of light and material from the sun. One solar flare may release as much energy as the whole world uses in 100,000 years.

Flares are short-lived, typically lasting 20 minutes. They occur near sunspots and seem to be energized by strong local magnetic fields.

Figure 7.13. Solar flare.

Solar flares produce gamma rays, X rays, ultraviolet radiation, radio waves, and particles, in addition to visible light. A large flare can hurl fantastic amounts of radiation and particles, as much as a billion exploding hydrogen bombs, into the solar system.

These reach Earth in a few minutes or days. A flare's radiation and particles could destroy all life on our planet if Earth were not shielded by its magnetic field and atmosphere. Astronauts in space must be protected, too.

When electrically charged particles from the sun strike Earth's atmosphere, they can stimulate the atmospheric atoms and ions to radiate light, producing auroras.

The *aurora borealis* (northern lights) and *aurora australis* (southern lights) are bright bands of light that sometimes appear in the night sky mainly in Earth's Arctic and Antarctic regions but occasionally also at middle latitudes in the United States. Maximum auroral activity occurs around 68° latitude. Auroras are visible about 2 days after a solar flare. They reach their peak about 2 years after sunspot maximum.

Large solar flares send out strong blasts of electrically charged particles that interact with Earth's magnetic field and upset it, causing magnetic storms. Compasses don't work normally then. The radiation and ions cause atmospheric storms, disrupt radio transmission, and cause surges on power lines as well.

Because solar flares affect Earth's environment so much, high priority is placed on predicting when they will occur. So far, no one can foretell exactly when a flare will happen. But the United States government carefully monitors the sun's magnetic field and activity daily.

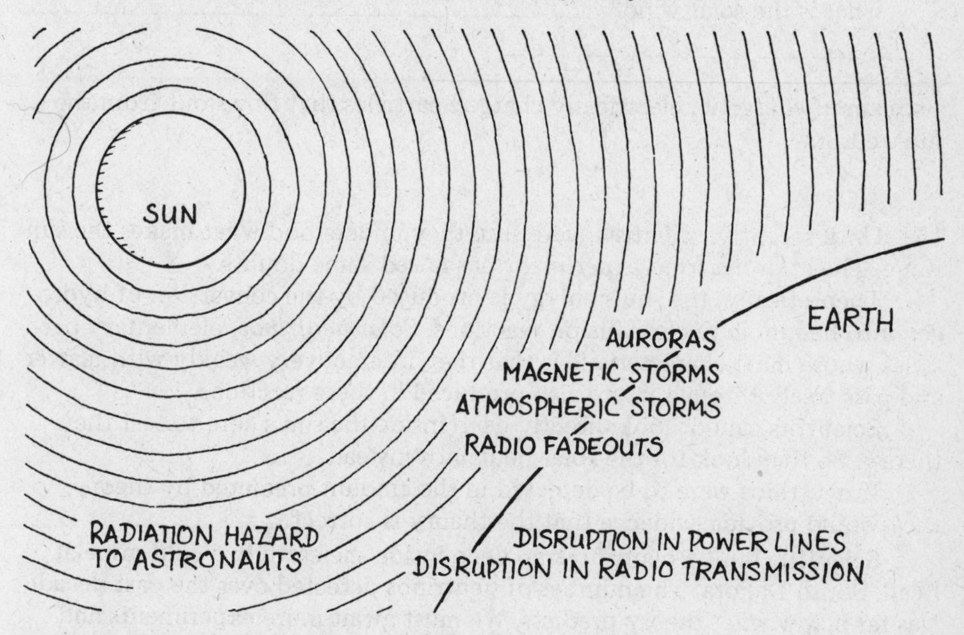

Figure 7.14. The effects of solar flares on Earth environment.

List two effects large solar flares have on modern technology on Earth.

---

(1) disruption in radio transmission; (2) disruption in power transmission

14. The *solar wind* is a stream of energetic, electrically charged particles that flows out from the sun's corona toward the edge of the solar system at all times.

The *average solar wind speed* near Earth is about 900,000 miles per hour (400 km/sec). Earth's atmosphere and magnetic field ordinarily protect us from harmful effects of the solar wind.

The solar wind is observed by instruments carried on spacecraft above Earth's atmosphere. In space, *wind temperatures* are thousands of degrees, and *wind velocities* are hundreds of thousands of miles per hour. They vary with solar activity.

Big blasts of solar wind occur during solar flares. The wind is strongest during periods when many sunspots are visible and solar activity is great. Strong blasts of solar wind can produce especially brilliant auroras.

The solar wind seems to come from regions in the sun's corona, called *coronal holes*, where gases are cooler and much thinner than elsewhere. The coronal holes are relatively weak and spread out, allowing high-speed solar wind streams to escape.

What is the solar wind? ______________________________

— — — — — — — — — — — — — — — — — —

a stream of energetic, electrically charged particles that flows out from the sun's corona

15. Until recently, scientists were sure they understood what makes the sun shine. The *solar neutrino experiment* has raised some doubts.

Theoretically, the sun's energy is produced by the conversion of hydrogen into helium in nuclear fusion reactions. *Solar neutrinos*, elementary particles whose main characteristic is that they interact very weakly with matter and pass freely through it, are also produced in these reactions.

Scientists cannot look directly deep inside the sun's core to test their theory. So they look for the solar neutrinos instead.

If neutrinos were to be detected in the amount predicted by theory, they would provide evidence that the theory is correct.

Scientists built a neutrino trap deep inside the earth in a gold mine at Lead, South Dakota. The number of neutrinos detected over the past decade was far below what theory predicts. We must await more experiments and analysis to solve this puzzle of the *missing neutrinos*.

Give two possible explanations for the unexpectedly small number of solar neutrinos detected so far. ______________________________

____

— — — — — — — — — — — — — — — — — —

(1) The experiment has been performed or interpreted incorrectly or incompletely.

(2) The theory of how the sun's energy is produced is incorrect or incomplete.

16. Write a short summary describing three phenomena that indicate violent activity on the sun, and name their probable cause. ____________________

____

____

____

— — — — — — — — — — — — — — — — — —

Your answer should briefly describe

(1) sunspots, or dark, relatively cool, temporary spots on the sun's photosphere;

(2) prominences, or relatively cool, dense gas that rises flamelike and bends over, above the sun's limb, in the hotter, thinner corona; and

(3) flares, or sudden, short-lived outbursts of light in the chromosphere near a sunspot.

Most violent activity on the sun seems to be caused and controlled by very strong local magnetic fields.

17. The sun is racing through space, just as all other stars do.

With respect to nearby stars, the sun is speeding toward the constellation Hercules at 45,000 miles per hour (20 km/sec), carrying its nine planets along.

Since the sun with its planets is inside the Milky Way Galaxy, it goes around our Galaxy's center as the whole Galaxy turns around in space. It travels at about 563,000 miles per hour (250 km/sec), as described in frame 5.2.

State two motions of the sun in the Milky Way Galaxy. ____________

____

— — — — — — — — — — — — — — — — — —

(1) toward the constellation Hercules; (2) around the Galaxy's center

SELF-TEST

This self-test is designed to show you whether or not you have mastered the material in Chapter Seven. Answer each question to the best of your ability. Correct answers and review instructions are given at the end of the test.

1. List three reasons why modern astronomers study the sun. ________

____

2. Match the most appropriate tool to the work:

___ (a) image processes in the hottest active regions of the sun

___ (b) photograph corona outside of solar eclipse

___ (c) photograph the sun's visible surface

___ (d) photograph the sun in the light of a particular element

___ (e) receive and record solar radio waves

1. coronagraph
2. optical solar telescope
3. radio telescopes
4. spectroheliograph
5. ultraviolet, X-, and gamma ray telescopes

3. Define the astronomical unit (AU). ________

____

4. Sketch the sun, and identify the corona, chromosphere, photosphere, zone of convection, and core.

5. Give the (a) radius, (b) mass, and (c) surface temperature of the sun.

____

6. Identify the following phenomena of the sun:

___ (a) on the bright photosphere: bright area surrounding a sunspot that appears on a spectroheliogram

___ (b) bright spots that look like rice grains on the photosphere

___ (c) dark, relatively cool blotches on the bright photosphere

___ (d) flamelike mass of bright gas rising hundreds of thousands of miles above the limb

___ (e) tremendous, short-lived, explosive outburst of light and material

1. flare
2. granules
3. plage
4. prominence
5. sunspots

7. Why is the sunspot cycle carefully monitored from Earth? ______________________________

8. What is the solar wind? ______________________________
9. List four ways that a flare and unusually big blasts of solar wind can affect Earth's environment. ______________________________

10. (a) What is the solar constant? (b) Why is it important to know if it is truly constant or varies with time? ______________________________

ANSWERS

Compare your answers to the questions on the self-test with the answers given below. If all of your answers are correct, you are ready to go on to the next chapter. If you missed any questions, review the frames indicated in parentheses following the answer. If you miss several questions, you should probably reread the entire chapter carefully.

1. (1) The sun is a free, nonpolluting, almost inexhaustible source of present and potential future energy.
(2) The sun is the only star close enough to observe in detail, so astronomers use it to determine what other stars are like.
(3) Changes in the sun's energy output affect Earth's climate, atmosphere, and weather, as well as power-transmission and communication systems. (frames 1, 13)

2. (a) 5; (b) 1; (c) 2; (d) 4; (e) 3 (frame 7)
3. The astronomical unit (AU) is the average *distance* between the Earth and the sun, about 93 million miles (officially 149,597, 870 km). (frame 2)

4.

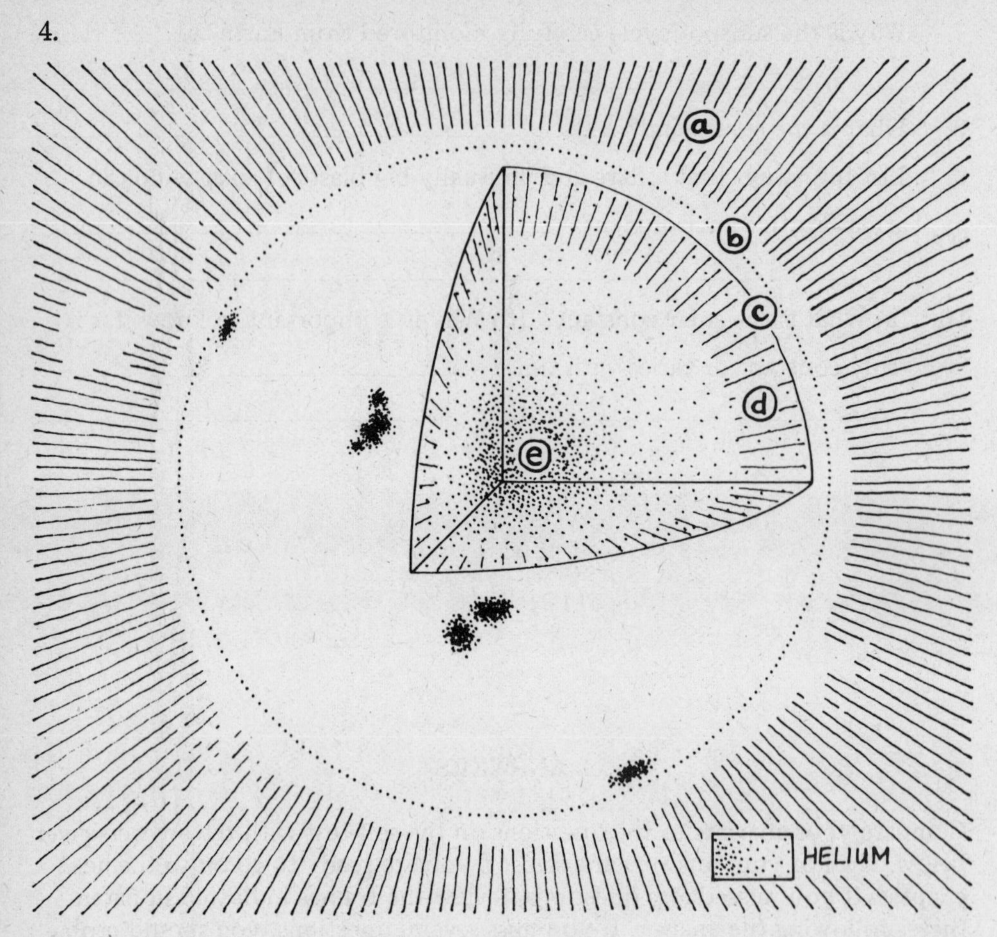

Figure 7.15. Regions of the sun. (a) corona (b) chromophere (c) photosphere (d) zone convection (e) core

(frame 4)

5. (a) 432,000 miles (696,000km); (b) about 2×10^{33} grams (1.989×10^{33} grams); (c) about 5,800° K (10,000° F)
(frames 2, 3, 4, 6)

6. (a) 3; (b) 2; (c) 5; (d) 4; (e) 1
(frames 8, 9, 12, 13)

7. The sunspot cycle is watched carefully from Earth as an indicator of solar activity. The sun is most active with greatest outbursts of energy and radiation during the years when sunspots are most numerous. It is least active in the years of sunspot minima. (frames 10, 13, 14)

8. a stream of energetic electrically charged particles that flows out from the sun's corona (frame 14)

9. (1) increased hazardous radiation; (2) auroras; (3) magnetic storms; (4) atmospheric storms (frames 13, 14)

10. (a) the amount of the sun's energy that falls per second on Earth's outer atmosphere, about 126 watts per square foot

    (b) Changes in the solar constant might drastically change Earth's climate and atmosphere. (frames 1, 13)

CHAPTER EIGHT
Understanding the Solar System

Finally we shall place the Sun himself at the center of the Universe.

Nicolaus Copernicus
De Revolutionibus Orbium Coelestium (1543)

1. Our *solar system* includes nine planets with their moons, thousands of minor planets called asteroids, comets, and dust particles that orbit the sun.

Planets are less massive and colder than stars. Stars generate their own light, but planets are not massive enough for nuclear fusion reactions to ignite. Planets shine by reflecting sunlight.

The planets of our solar system range in mass from lightest Pluto to heaviest Jupiter—318 times Earth's mass. The whole planetary system has less than $\frac{1}{740}$ the mass of the sun.

In size, the planets range from smallest Pluto to largest Jupiter, which is 89,000 miles (142,700 km) in diameter. Giant Jupiter's diameter is less than $\frac{1}{9}$ of the sun's.

Figure 8.1. Relative sizes of the sun, planets, and Earth's moon.

What is the essential difference between a planet and a star? ______________

____

- - - - - - - - - - - - - - - - - - - -

A planet is less massive and colder than a star. It shines by reflecting light from a star. (A star generates its own light.)

2. All of the planets *revolve*, or travel around, the sun in the same direction from west to east. This movement is called *direct motion.*

The plane of the Earth's orbit around the sun is called the *ecliptic plane.* The orbits of all the other planets except Pluto are in nearly the same plane, like the grooves in a phonograph record. These facts support the condensation theory of the origin of the solar system (frame 7.3). All the planets and the sun probably formed together out of an eastward rotating cloud (figure 7.3).

The planets that follow orbits closer to the sun than Earth's are called *inferior*, while those with orbits outside Earth's are called *superior* planets.

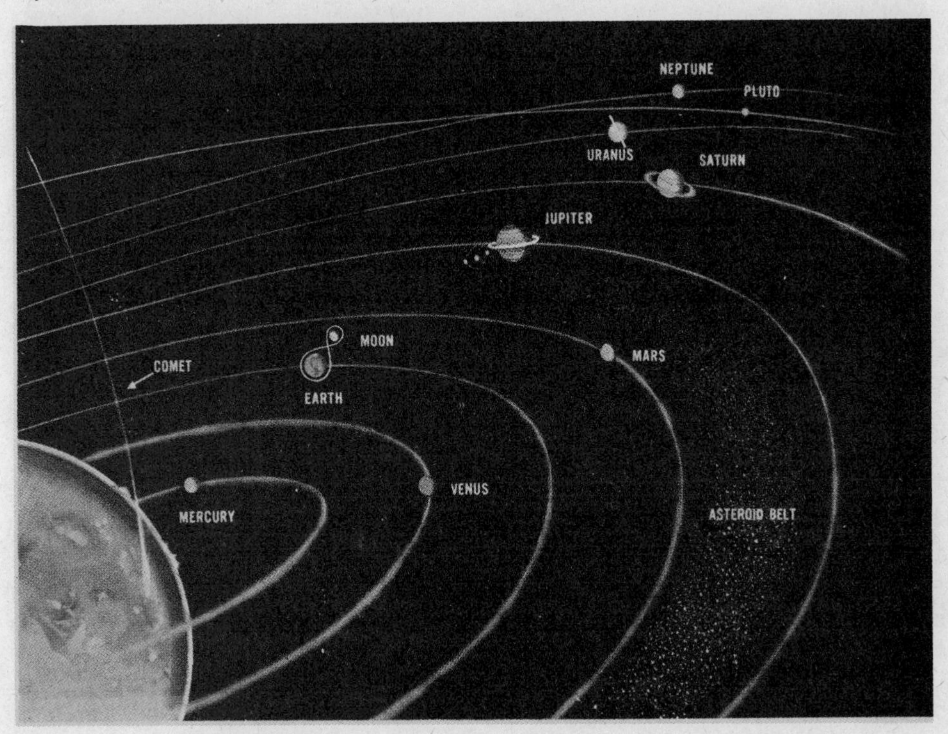

Figure 8.2. The inferior and superior planets.

Look at figure 8.2. List the (a) inferior and (b) superior planets.

(a) ____

(b) ____

- - - - - - - - - - - - - - - - - - - -

(a) Mercury, Venus; (b) Mars, Jupiter, Saturn, Uranus, Neptune, Pluto

3. Five planets—Mercury, Venus, Mars, Jupiter, and Saturn—look like very bright stars in the sky. The sun, moon, and these five bright planets were known to the ancients. Each was thought to rule one day of the week, which was given its name (in Latin).

We take the names for our days of the week from the Anglo-Saxons, who substituted the names of equivalent gods and goddesses for the Roman ones. French and Spanish names are adapted directly from the Latin.

Day	Ruling Planet	Anglo-Saxon Equivalent	Latin	French	Spanish
Sunday	Sun	—	Dies Solis	Dimanche	Domingo
Monday	Moon	—	Dies Lunae	Lundi	Lunes
Tuesday	Mars	Tiw	Dies Martis	Mardi	Martes
Wednesday	Mercury	Woden	Dies Mercurii	Mercredi	Miercoles
Thursday	Jupiter (Jove)	Thor	Dies Jovis	Jeudi	Jueves
Friday	Venus	Freya	Dies Veneris	Vendredi	Viernes
Saturday	Saturn	—	Dies Saturni	Samedi	Sabado

(a) Which of our days of the week is closest to the original Latin god's name? ______________________

(b) Which days of the week carry the names of Anglo-Saxon gods?

____

(a) Saturday; (b) Tuesday, Wednesday, Thursday, Friday

4. The moon is Earth's only natural *satellite*, circling our planet as we travel around the sun. It *shines* in the sky by *reflecting* sunlight.

The moon's appearance changes regularly every month. Half of the moon is always lighted by the sun, but the bright shape we see from Earth, called its *phase*, changes as the moon travels around our planet.

Refer to figure 8.3. The *new moon* is dark. It is not seen in the sky because the moon's night side is facing Earth. The *waxing* (growing bigger) *crescent* moon follows a few days later. You often see its disk faintly lighted by sunlight reflected back off Earth, called *earthshine*.

About 7 days after the new moon, when the moon has traveled $\frac{1}{4}$ of its way around Earth, it rises around noon, and we observe *first quarter* shine. The *waxing gibbous* moon follows, with more than half of the moon's bright disk shining toward Earth.

When the moon is about two weeks into its cycle, the *full moon* lights the sky all night with its whole, bright disk. The visible part of the moon's bright disk then *wanes*, or decreases in size, as the moon completes its trip around Earth during the last two weeks of its cycle.

The time required for the moon's phases to repeat, called the *synodic month*, is $29\frac{1}{2}$ days ($29^d 12^h 44^m 2^s.9$).

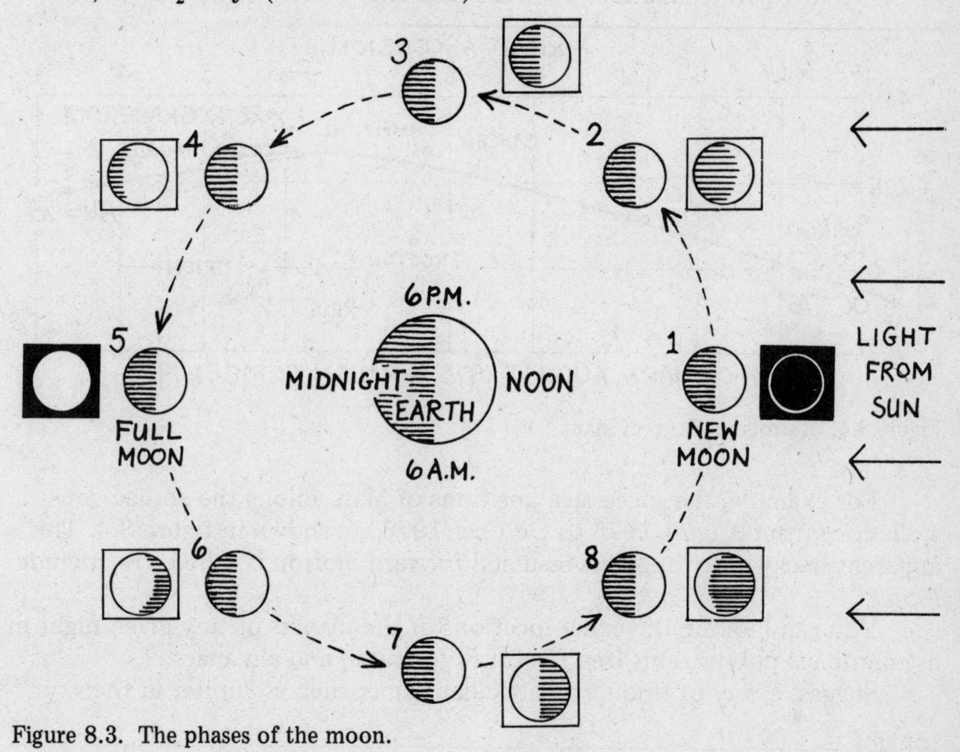

Figure 8.3. The phases of the moon.

Identify the phase of the moon corresponding to the indicated position in its orbit for each of the following phases, as labeled in figure 8.3. (a) waxing crescent; (b) first quarter; (c) waxing gibbous; (d) waning gibbous; (e) third quarter; (f) waning crescent

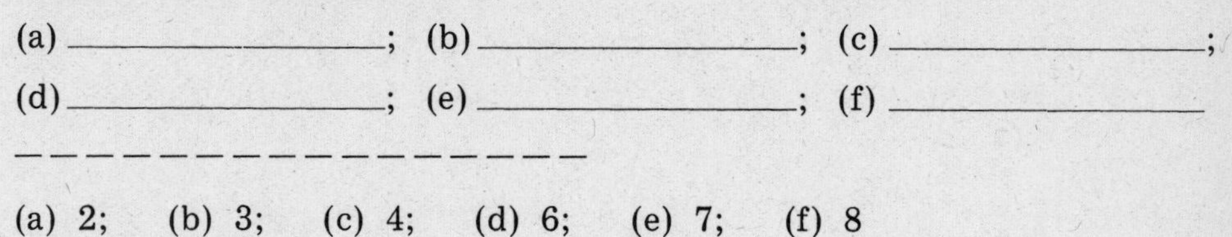

(a) ________________; (b) ________________; (c) ________________;
(d) ________________; (e) ________________; (f) ________________

— — — — — — — — — — — — — — — — —

(a) 2; (b) 3; (c) 4; (d) 6; (e) 7; (f) 8

5. *Planet positions* are not marked on star maps. Stars keep their same relative positions in the sky for decades, but planets do not. The word "planet" comes from the Greek for "wanderer." Planets move across the celestial sphere near the ecliptic.

Venus and Mercury appear to move forwards and backwards on either side of the sun in Earth's sky. Maximum *elongation*, or maximum distance east or west of the sun, is 48° for Venus and 28° for Mercury.

Mars, Jupiter, and Saturn wander generally eastward through the constellations of the zodiac. At times planets seem to turn and move in reverse, or *retrograde*, before resuming forward motion.

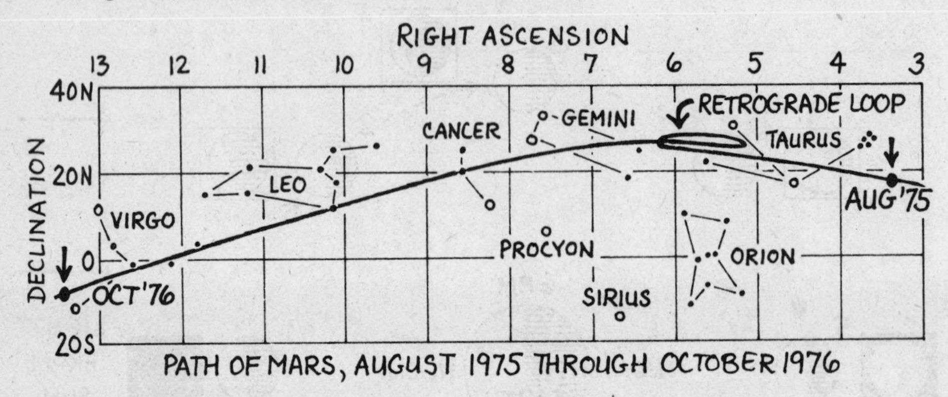

Figure 8.4. Retrograde loop of Mars.

For example, the successive positions of Mars among the zodiac constellations from August 1975 to October 1976 are shown in figure 8.4. The apparent backward swing plus resumed forward motion is called a *retrograde loop*.

You can look up the exact locations of the planets on any given night in astronomical publications (see Useful References) and almanacs.

Suggest a way of finding a particular planet such as Jupiter in the sky tonight. ______________________________

____

____

- - - - - - - - - - - - - - - - - -

Find out which zodiac constellation Jupiter is in today by using an astronomical publication or almanac. Locate that constellation on your star maps. For example, suppose you find Jupiter is located in Taurus today. If Taurus is in the sky tonight you can easily spot Jupiter. It will be the brilliant "star" that does not belong to the constellation.

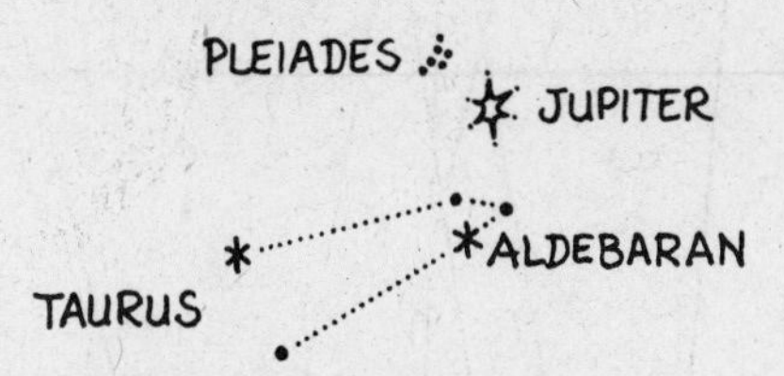

Figure 8.5. Jupiter in the constellation Taurus.

★ Keep a record of the position of Venus and Mars for several months. Observe them in the sky if possible. Use the information in this chapter to explain the motions you observe.

6. The search for a simple explanation of the planets' observed motions in the sky changed mankind's view of the world.

In *Almagest*, written about 150 A.D., the Alexandrian astronomer Ptolemy described the ancient *geocentric*, or Earth-centered, view of the universe. Circles were considered "perfect" shapes. The sun, moon, and planets were supposed to move on small circles called *epicycles*, whose centers moved around Earth on larger circles called *deferents*.

For more than fourteen centuries, this *Ptolemaic system* was accepted as the basis for astronomical work. It described with considerable accuracy the observed positions and motions of the heavenly bodies known at that time. And it expressed the commonsense view of the world that people had from observing the sky. With minor modifications, this Earth-centered theory became part of the dogma of the Roman Catholic Church of the Middle Ages.

Nicolaus Copernicus (1473-1543), a Polish astronomer, published his radical *heliocentric*, or sun-centered, theory the year he died. In the *Copernican system*, the planets, including Earth, circle around a stationary central sun. According to the Copernican theory, the apparent wandering motions of the planets result from a combination of the real orbital motions of both Earth and observed planets.

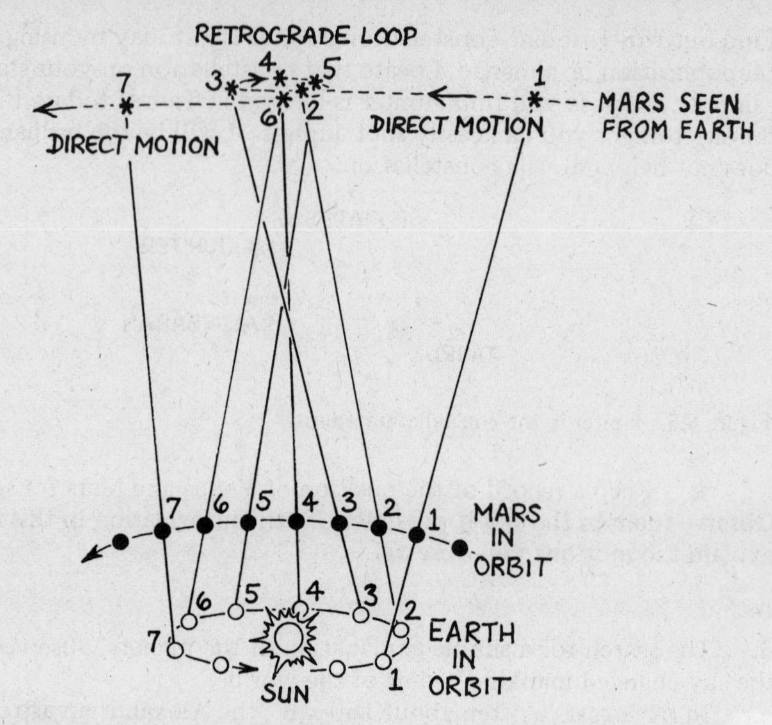

Figure 8.6. The apparent motion of Mars explained.

The apparent path of Mars, illustrated in Figure 8.4, is explained as follows. Mars never really moves backward in its orbit. Planets always go forward. The retrograde loop in the sky is caused by the relative motion of Earth and Mars. Our faster-moving Earth catches up to Mars and moves past it. So Mars, the outer planet, looks to us as if it is moving backward. (It is like watching a slower car while you are overtaking it in a faster car.)

What change in the philosophical view of Earth was required for people to accept the Copernican theory instead of the Ptolemaic? ______________

____

- - - - - - - - - - - - - - - - - - - -

Earth could no longer be considered the center of the entire universe and supremely important.

7. The Italian scientist Galileo Galilei (1564-1642) provided important observational data for the Copernican system. Galileo was the first person to use a telescope for sky observations. He observed that Venus appears to change its shape regularly.

The Ptolemaic system could not account for the phases of Venus. But the Copernican system had a simple explanation. Venus and Mercury, the two inferior planets, show phases as they reflect sunlight to Earth from different places in their orbits around the sun.

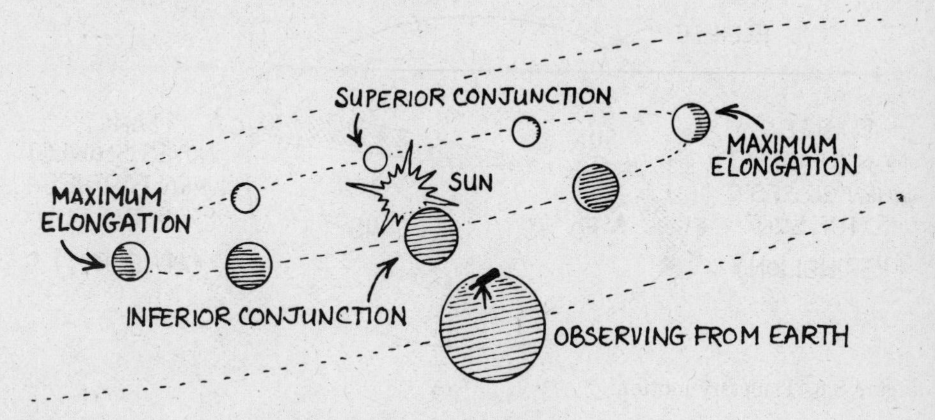

Figure 8.7. The phases of Venus as viewed from Earth.

When he was almost 70 years old, after a brilliant career, Galileo was charged by the Roman Catholic Church with believing and holding doctrines contrary to what it considered Divine Truth. He was forced to renounce his view that the Copernican system was correct.

Explain why Galileo's observation of the phases of Venus in his small telescope provided important support for the Copernican (heliocentric) view of the universe. ____

____

— — — — — — — — — — — — — — — — —

If Venus and the sun circled Earth, as the geocentric theory said, Venus would not show phases. Venus must move around the sun for us to see different parts of its daylight side (phases) at different times.

8. The German astronomer *Johannes Kepler* (1571-1630) deduced a simple, precise description of planetary motion. He worked from records of almost 20 years of sky observations inherited from *Tycho Brahe* (1546-1601), a Danish astronomer.

Kepler's laws of planetary motion greatly improved the accuracy of predictions of planet positions. The three laws state:

1. Each planet moves around the sun in an orbit that is an *ellipse* with the sun at one focus.

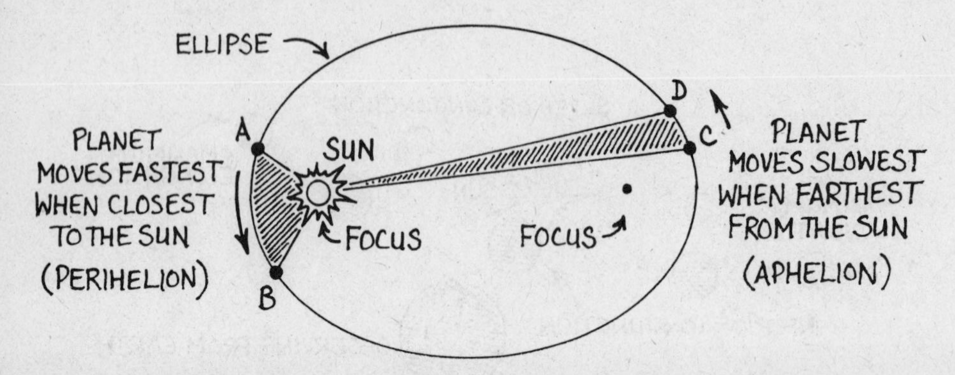

Figure 8.8. Planetary motion.

2. Each planet moves so that an imaginary line joining the sun and the planet sweeps out equal areas in equal times. The planet goes from A to B and from C to D in the same amount of time. In other words, planets move fastest when they are closest to the sun (*perihelion*) and slowest when they are farthest away (*aphelion*).

3. The squares of the period of time required for any two planets to complete a trip around the sun have the same ratio as the cubes of their average distance from the sun.

Kepler's third law can be used to find a planet's distance, d, from the sun compared to Earth's distance of 1 AU (frame 7.2). The planet's orbital period, p, in years is found from observations. Then Kepler's third law is written $d^3 = p^2$. For example, Jupiter's orbit period is 11.86 years. So Jupiter's distance, d, from the sun is found from $d^3 = (11.86)^2 \cong 141$. Solving this for $d = \sqrt[3]{141}$, we find $d = 5.2$ AU.

How far would a planet be from the sun if its orbital period were observed to be 8 years? Explain. ______________________________

____

---

4 AU. According to Kepler's third law, $d^3 = p^2$. So $d^3 = (8)^2 = 64$. And $d = \sqrt[3]{64} = 4$.

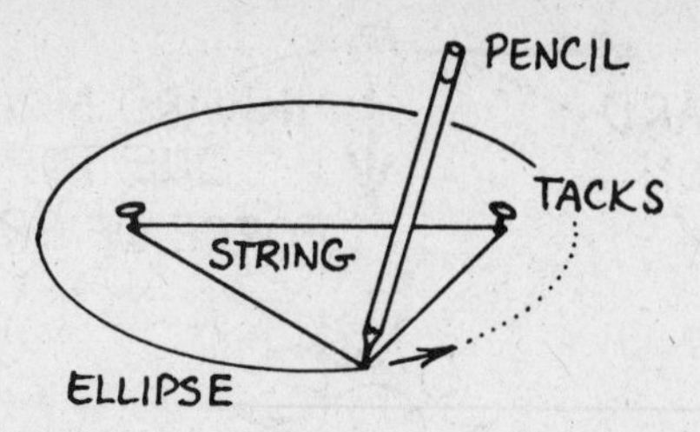

Figure 8.9. Drawing an ellipse.

To draw an ellipse, put two tacks in a board. Tie a string around them. Trace the ellipse by keeping the string taut with your pencil point.

9. Kepler's laws explain how the planets are *observed* to move. The English mathematician and natural philosopher Sir *Isaac Newton* (1642-1727) formulated the law that explains *why* they move as they do.

Newton's *law of gravity* states that any two objects of masses m_1 and m_2 separated from each other by a distance, d, attract each other with a force (F), called gravity, that is directly proportional to the product of their masses and inversely proportional to the square of their distance away from each other.

The formula is:

$$F = G \frac{m_1 m_2}{d^2},$$

where G = gravitational constant

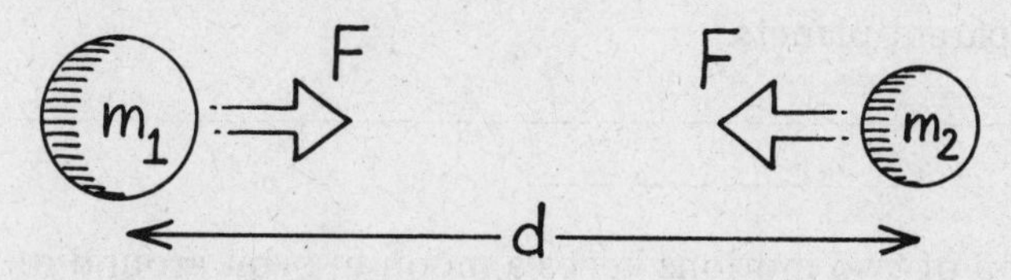

Figure 8.10. Newton's law of gravity.

A force of attraction is necessary to keep the planets moving in their curved paths around the sun. Without this force they would move straight away into space. The required force is provided by the sun's gravity which continually pulls the planets in toward the sun.

The combination of their forward motion and their motion in toward the sun under gravity keeps the planets travelling in their orbits around the sun.

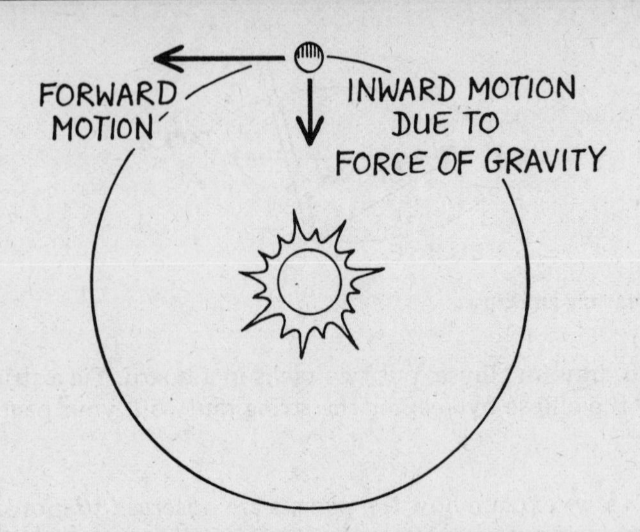

Figure 8.11. The two motions that keep a planet in orbit.

Newton's genius was to realize that his law of gravitation applies to falling objects on Earth, to the motion of the moon and planets, and to all material bodies. His law is called the *universal law of gravitation*, meaning that it is true for all objects everywhere in the universe.

Newton proposed that the law of gravitation and the three laws of motion that he formulated were basic laws of physics. He generalized and mathematically derived Kepler's laws of planetary motion from basic principles. He invented and used in his work the branch of mathematics we call calculus.

Apply Newton's laws to explain why satellites (moons) stay in orbit around their parent planets.

- - - - - - - - - - - - - - - - - -

A combination of two motions keeps a moon in orbit around its parent planet—its forward motion and its inward motion caused by the pull of the planet's gravity.

10. Our moon travels around Earth at an *average speed* of 2,295 miles per hour (1.02 km/sec).

The time for the moon to make one complete trip around Earth with respect to the stars, about $27\frac{1}{3}$ days ($27^{d}7^{h}43^{m}11^{s}.5$), is called a *sidereal month.*

The moon's *average angular diameter* in the sky is about $\frac{1}{2}°$ (31′5″ of arc). Since the moon's path around Earth is an *ellipse*, the moon looks larger at *perigee* (the point on the orbit closest to Earth) and smaller at *apogee* (the point on the orbit farthest from Earth).

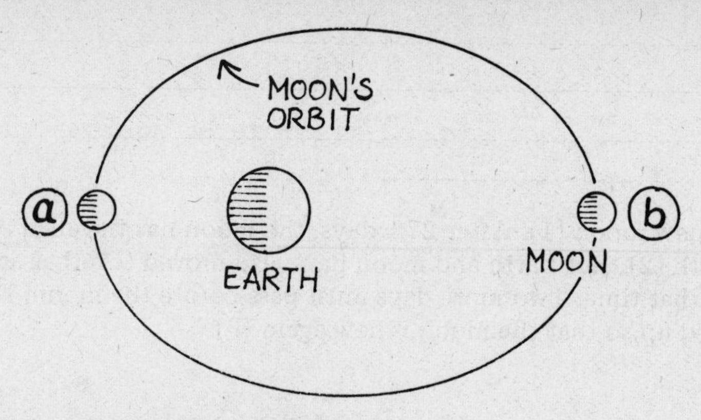

Figure 8.12. The moon's orbit around Earth.

Identify the apogee and perigee in figure 8.12, and indicate where on the ellipse the moon looks larger and where it looks smaller than average.

(a) ______________________; (b) ______________________

(a) perigee; the moon looks larger (b) apogee; the moon looks smaller

11. If you enjoy word games or crossword puzzles, you'll find syzygy a good word to know. It means three celestial bodies in a line, such as sun-moon-Earth.

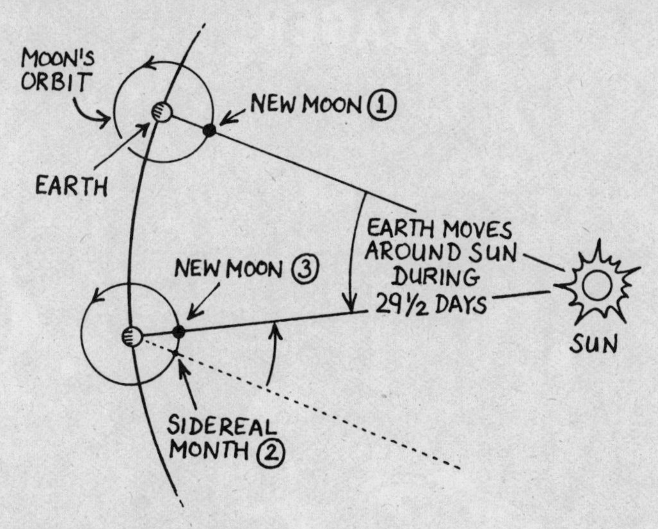

Figure 8.13. The moon's sidereal month.

Refer to figure 8.13. Explain why the moon's synodic month, or the month of the moon's phases, is 2 days longer than its sidereal month (see frame 4). ____

____

____

— — — — — — — — — — — — — — — — — —

Start with new moon (1). After $27\frac{1}{3}$ days, the moon has travelled completely around Earth (2). But Earth and moon have also moved together around the sun during that time. Two more days must pass before the moon, Earth, and sun are lined up so that the moon is new again (3).

12. Spacecraft obey the same basic laws of physics that natural astronomical bodies do.

Earth-orbiting spacecraft are called *artificial satellites.* Rockets launch these satellites into orbit with a forward velocity. The combination of their forward motion and their motion in toward the Earth under Earth's gravity keeps them in their orbits.

Unmanned spacecraft are sent to explore the planets. These are launched with a forward velocity into orbit around the sun. Their motions are calculated using Newton's laws, just as planetary motions are. The most ambitious multi-targeted space flight yet is Project Voyager, illustrated in figure 8.14.

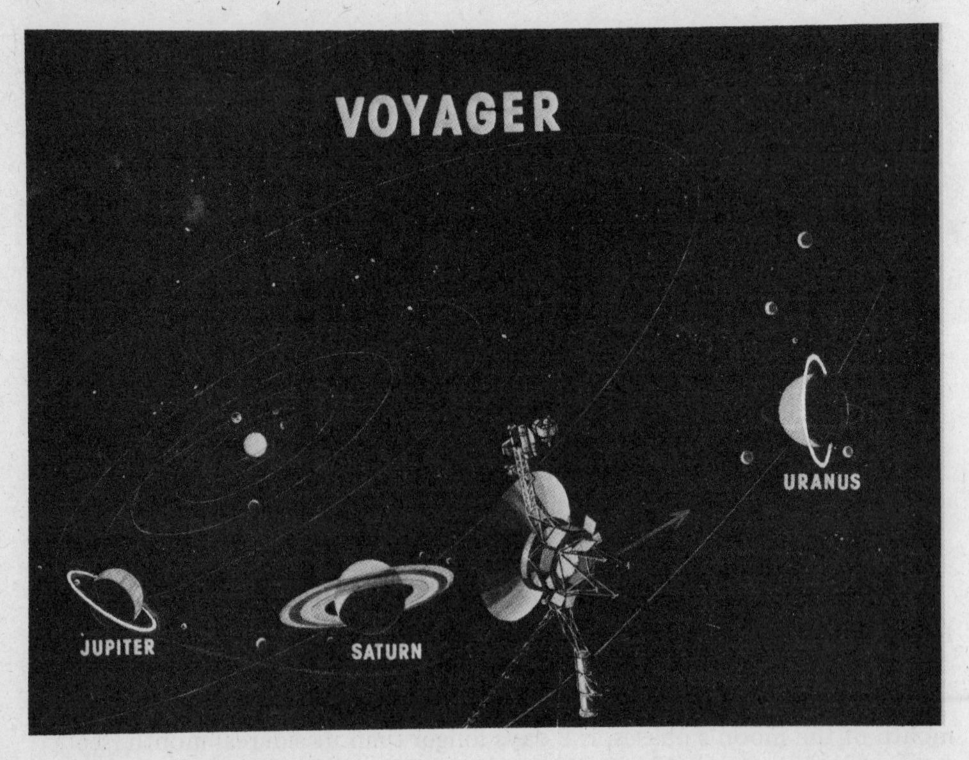

Figure 8.14. Project Voyager.

Its 12-year timetable from launch in 1977 included encounters with Jupiter and Saturn and several moons of both planets, Uranus, and possibly Neptune.

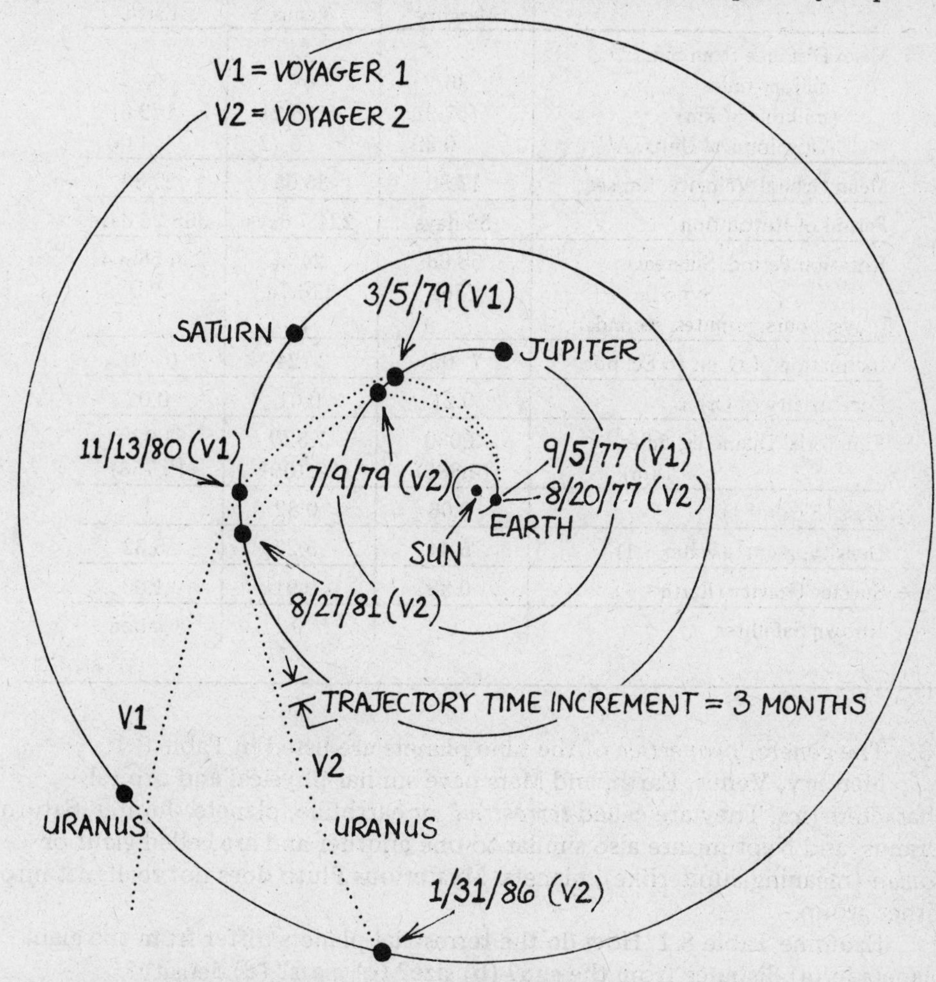

Figure 8.15. Planned flight path of two Voyager spacecraft.

Figure 8–15 shows the planned flight path of the two Voyager spacecraft scheduled for the entire mission. Spacecraft and planetary paths are marked to indicate progress. Dates show the down-to-the-minute planning that is done to pre-aim the spacecraft instruments for visits to several planets.

Refer to figure 8.15. About how many years after launch did Voyager 2 reach (a) Jupiter? __________ (b) Saturn? __________

After how many years of travel will it reach (c) Uranus? __________

(d) Neptune? __________

(a) 2 years; (b) 4 years; (c) 9 years; (d) 12 years

Table 8.1. Properties of the planets.

	Mercury	Venus	Earth
Mean Distance from Sun			
millions miles	36	67	93
(millions of km)	(57.9)	(108.2)	(149.6)
Astronomical Units, AU	0.39	0.72	1.00
Mean Orbital Velocity, km/sec	47.90	35.05	29.80
Period of Revolution	88 days	224.7 days	365.26 days
Rotation Period, Sidereal	58.6d	243d	23h 56m 4s
Synodic	176d	116.7d	24h
(days, hours, minutes, seconds)			
Inclination of Orbit to Ecliptic	7° 00′	3° 24′	0° 00′
Eccentricity of Orbit	0.21	0.01	0.02
Equatorial Diameter, miles	3,030	7,520	7,930
(km)	(4,880)	(12,100)	(12,756)
Mass (Earth = 1)	0.06	0.82	1
Density, g/cm^3 (Water = 1)	5.44	5.27	5.52
Surface Gravity (Earth = 1)	0.38	0.91	1.00
Known Satellites	0	0	1 moon

13. The *general properties* of the nine planets are listed in Table 8.1.

Mercury, Venus, Earth, and Mars have similar physical and orbital characteristics. They are called *terrestrial*, or earthlike, planets. Jupiter, Saturn, Uranus, and Neptune are also similar to one another and are called *giant* or *Jovian* (meaning Jupiterlike), planets. Mysterious Pluto does not really fit into either group.

Examine Table 8.1. How do the terrestrial planets differ from the giant planets in (a) distance from the sun? (b) size? (c) mass? (d) density?

	Terrestrial Planets	Giant Planets
(a)	__________	__________
(b)	__________	__________
(c)	__________	__________
(d)	__________	__________

- - - - - - - - - - - - - - - -

	Terrestrial Planets	Giant Planets
(a)	near the sun	far from the sun
(b)	small diameter	large diameter
(c)	small mass	large mass
(d)	high density	low density

Table 8.1. Properties of the planets. (Continued)

Mars	Jupiter	Saturn	Uranus	Neptune	Pluto
142 (228.0) 1.52	484 (778.4) 5.20	885 (1,424.6) 9.52	1,780 (2,867) 19.16	2,790 (4,486) 29.99	3,660 (5,890 39.37
24.14	13.06	9.65	6.80	5.43	4.74
687 days	11.86 years	29.46 years	84.01 years	164.1 years	247 years
24h 37m 23s 24h 39m 35s	9h 55m 30s 9h 55m 33s	10h 39m 26s 10h 39m 26s	~16h? ~16h?	~18½h? ~18½h?	6d 9h? 6d 9h?
1° 51′	1° 18′	2° 29′	0° 46′	1° 47′	17° 10′
0.09	0.05	0.05	0.05	0.01	0.25
4,220 (6,794)	89,000 (143,200)	75,000 (120,000)	32,200 (51,800)	30,800 (49,500)	~1,620? (~2,600?)
0.11	317.9	95.2	14.6	17.2	<0.1?
3.95	1.31	0.70	1.21	1.66	0.8?
0.38	2.34	0.93	0.85	1.14	0.04?
2 moons	16 moons Rings	22 moons Rings	5 moons Rings	2 moons	1 moon

14. Notice in Table 8.1 the difference in the length of *days* and *years* on the terrestrial and giant planets.

The *period of revolution* is the length of time required for a planet to complete one trip around the sun. The period of revolution is the length of one year on that planet. A year on the terrestrial planets ranges from 88 (Mercury) to 687 earth-days (Mars). A year on the giant planets is much longer, equal to many earth-years.

Planets all *rotate*, or turn around on an axis, as they revolve around the sun. The *period of rotation* is the length of time required for the planet to turn completely around once. The sidereal rotation period is the length of one sidereal day on the planet (review frame 1.23, if necessary).

A sidereal day is long on the terrestrial planets, because terrestrial planets rotate relatively slowly. It is short on giant planets, because they rotate rapidly. The shortest sidereal day (Jupiter) is only 9 earth-hours and 50 minutes long.

A planet's synodic rotation period is the length of one solar day on the planet, or the time interval between two successive meridian transits of the sun as would be seen by an observer on the planet.

Examine Table 8.1. (a) Which giant planet has the longest year, and how many earth-years long is it? ____________________

(b) Which terrestrial planet has the longest sidereal day, and how many earth-days is it? ______________________________

(c) Which planet has the longest solar day, and how many earth-days is it?

____

— — — — — — — — — — — — — — — — — —

(a) Neptune. It is equal to 164.1 earth-years.
(b) Venus. It is equal to 243 earth-days.
(c) Mercury. It is equal to 176 earth-days.

15. Refer to Table 8.1. (a) How many known satellites do terrestrial planets have altogether? ______________ (b) How many known satellites do the giant planets have? ______________ (c) Can you suggest a possible reason for this difference? ______________

____

— — — — — — — — — — — — — — — — — —

(a) Terrestrial planets have only three known moons—Earth has one; Mars has two.

(b) Giant planets have many moons and rings; forty moons and three ring systems have been discovered so far.

(c) Giant planets are much more massive with stronger gravity than terrestrial planets. Hence they could more readily hold moons that formed nearby or capture others that passed by.

16. A gap occurs between the orbits of Mars and Jupiter in which thousands of small, irregularly shaped, rocky bodies called minor planets, or *asteroids*, orbit the sun. Most follow paths inside this region called the *asteroid belt*.

Through a telescope, asteroids (means "starlike") look like stars. More than 2,000 asteroids have been catalogued, and millions more small ones probably exist. The largest and first sighted is 1 Ceres, 634 miles (1,020 km) across, discovered by Sicilian astronomer Guiseppe Piazzi (1746–1826) in 1801. The brightest is 4 Vesta, 341 miles (549 km) in diameter.

The total mass of all the asteroids together is probably less than 3 percent of our moon's. Asteroids are typically rocky and metallic in composition, with some chemical compounds of carbon, nitrogen, and oxygen plus hydrogen. Water (in the form of water of hydration) was first detected on Ceres.

The bright asteroids are probably clumps of mass that condensed from the original solar nebula but never got big enough to form a large planet. The fainter ones are probably fragments resulting from collisions.

Some asteroids cross the orbit of Mars and pass inside Earth's orbit. These are called *Apollo asteroids*. Apollo asteroids have come within a million miles of Earth. When a new Apollo asteroid is sighted, some people fear a disastrous collision. This very unlikely event has never been recorded.

A mysterious object named *Chiron* orbits the sun between Saturn and Uranus. Chiron's origin is uncertain. It may have been gravitationally tugged out of the asteroid belt. Chiron could be the first asteroid sighted in an outer zone. Or, it could be the remains of a comet or an escaped moon.

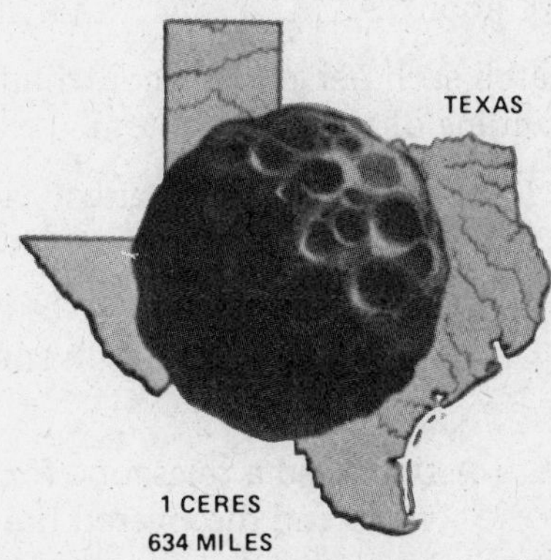

Figure 8.16. Sizes of famous asteroids: Largest Ceres, brightest Vesta, and Apollo asteroid Eros.

What are asteroids? ______________________________

____

swarms of irregular, rocky bodies that orbit the sun, mostly between the orbits of Mars and Jupiter

SELF-TEST

This self-test is designed to show you whether or not you have mastered the material in Chapter Eight. Answer each question to the best of your ability. Correct answers and review instructions are given at the end of the test.

1. List the members of the solar system. ____________________

____

2. What is the essential difference between a star and a planet? ____________

____

3. Give two facts that support the condensation theory of the formation of the solar system. ____________________

____

4. Which phase of the moon would you see if the moon were rising in the sky about (a) 6 P.M.? ____________________;

(b) noon? ____________________

5. Match each person to a contribution to the development of our understanding of the solar system.

___ (a) described a geocentric view of the universe in the *Almagest* about 150 A.D.

___ (b) determined his three laws of planetary motion empirically from observational data

___ (c) first used a telescope for astronomical work and discovered the phases of Venus

___ (d) wrote a book describing a heliocentric theory of planetary motions which was published in 1543, the year he died

___ (e) formulated the three fundamental laws of motion and the universal law of gravitation

___ (f) observed and recorded planetary motions for almost 20 years

1. Copernicus
2. Galileo
3. Kepler
4. Newton
5. Ptolemy
6. Tycho Brahe

6. What keeps planets in their orbits around the sun? ____________________

____

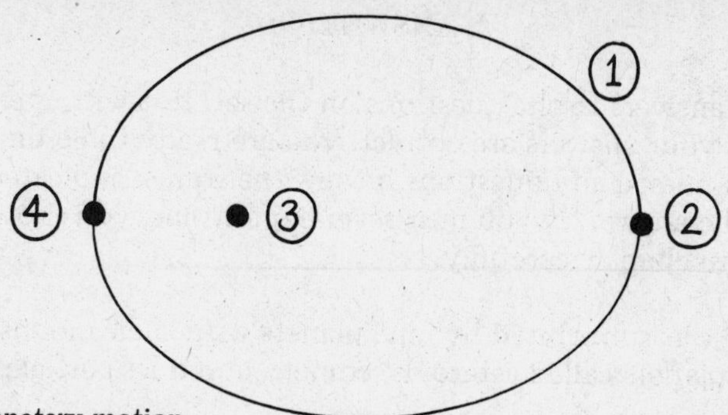

Figure 8.17. Planetary motion.

7. Referring to figure 8.17, identify the following points: (a) sun ______;
(b) ellipse ______; (c) aphelion ______; (d) perihelion ______;
(e) force of gravity is greatest ______; (f) planet moves slowest ______

8. By how much do the moon's sidereal month and synodic month differ? Explain why. ______

______

9. What force keeps the Voyager 1 and 2 spacecraft in their trajectories as they travel through the solar system?

______

10. Classify each of the following as a property of (1) terrestrial planets or (2) giant planets.

___ (a) far from the sun
___ (b) small diameter
___ (c) large mass
___ (d) low density
___ (e) short period of revolution
___ (f) short period of rotation
___ (g) many moons

11. Match a planet to the appropriate description. Hint: Refer to Table 8.1.

___ (a) closest to sun
___ (b) orbit most inclined to the ecliptic plane
___ (c) has the longest sidereal day
___ (d) has a year approximately equal to 2 earth-years
___ (e) most massive
___ (f) most dense

1. Mercury
2. Venus
3. Earth
4. Mars
5. Jupiter
6. Saturn
7. Uranus
8. Neptune
9. Pluto

12. What are asteroids? ______

______

ANSWERS

Compare your answers to the questions on the self-test with the answers given below. If all of your answers are correct, you are ready to go on to the next chapter. If you missed any questions, review the frames indicated in parentheses following the answer. If you miss several questions, you should probably reread the entire chapter carefully.

1. One star—our sun orbited by nine planets with their moons, thousands of minor planets called asteroids, comets, and dust particles. (frame 1)
2. mass. A planet is less massive and colder than a star. While a star generates its own light, a planet shines by reflecting light from a star. (frame 1)
3. All of the planets revolve around the sun in the same direction. The orbits of all the planets except Pluto lie nearly in the ecliptic plane. (frame 2)
4. (a) full moon; (b) first quarter (frame 4)
5. (a) 5; (b) 3; (c) 2; (d) 1; (e) 4; (f) 6 (frames 6, 7, 8, 9)
6. a combination of their forward motion and their motion in toward the sun under the sun's gravity (frame 9)
7. (a) 3; (b) 1; (c) 2; (d) 4; (e) 4; (f) 2 (frames 8, 9, 10)
8. two days. While the moon revolves around Earth, both Earth and the moon revolve together around the sun. (frames 10, 11)
9. gravity (frame 12)
10. (a) 2; (b) 1; (c) 2; (d) 2; (e) 1; (f) 2; (g) 2 (frames 13, 14, 15; Table 8.1)
11. (a) 1; (b) 9; (c) 2; (d) 4; (e) 5; (f) 3 (frames 13, 14; Table 8.1)
12. irregular, rocky bodies that orbit the sun, mostly between the orbits of Mars and Jupiter (frames 1, 16)

CHAPTER NINE
The Planets

The earth is the cradle of mankind, but one does not live in the cradle forever.

Konstantin Tsiolkovsky

1. *Mercury*, the planet closest to the sun, is often hidden in its glare. Mercury was appropriately named for the swift Roman messenger god. It wings around the sun fastest of all the planets, at 108,000 miles per hour (47.9 km/sec).

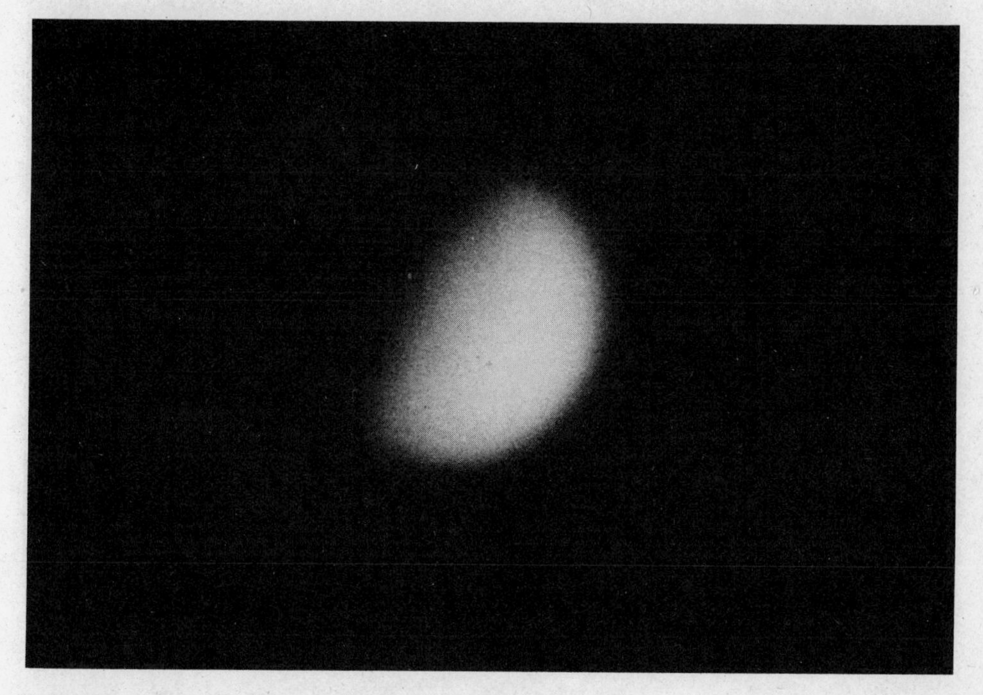

Figure 9.1. Mercury as photographed from Earth.

Figure 9.1 shows the best type of picture of Mercury that can be taken from Earth. Our first close-up look at Mercury came through the robot "eyes" of Mariner 10 which flew by within 436 miles (703 km) above Mercury's rugged surface in 1974 and 1975. In figure 9.2, the largest craters are about 200 km in diameter.

Figure 9.2. Mercury—a composite of pictures taken by Mariner 10.

Mercury looks like our moon with its many craters, mountains, and basins. Temperatures vary from extremely hot at noon—up to 800° F (700° K)—to bitter cold—down to -300° F (90° K) at midnight.

Mercury has a trace of an atmosphere. Helium, argon, oxygen, carbon, and xenon have been detected. The surface air pressure is barely 2 trillionths of ours at sea level ($< 2 \times 10^{-9}$ millibars). It also has a very weak magnetic field that affects the moving charged particles in the solar wind.

Mercury's craters suggest that meteorites bombarded the inner planets in the final stages of their formation. The largest, Caloris Basin, is 930 miles (1,300 km) across. Large smooth areas resembling the moon's maria (frame 10.8) suggest that extensive lava flooding occurred in the past.

If these flooded surface areas have a silicate composition like our moon's (average density about three times that of water), how could you explain Mercury's average density of almost $5\frac{1}{2}$ times that of water?

- - - - - - - - - - - - - - - - - -

Mercury probably has a very dense core. (Scientists suggest that Mercury has a dense iron core.)

2. Brilliant *Venus* was named for the Roman goddess of love and beauty. At night Venus outshines all the stars. The planet is so conspicuous that it is often mistakenly reported as an unidentified flying object (UFO).

Venus, like Mercury, circles the sun inside Earth's orbit. As a result, both planets go through a cycle of phases (figure 8.7) that are visible in a small telescope. Venus rotates from east to west, or *retrograde*.

Venus and Mercury shine in the western sky just after sunset near their eastern elongation. Then they appear to *follow* the sun across the sky. They are frequently called *evening stars* at that time.

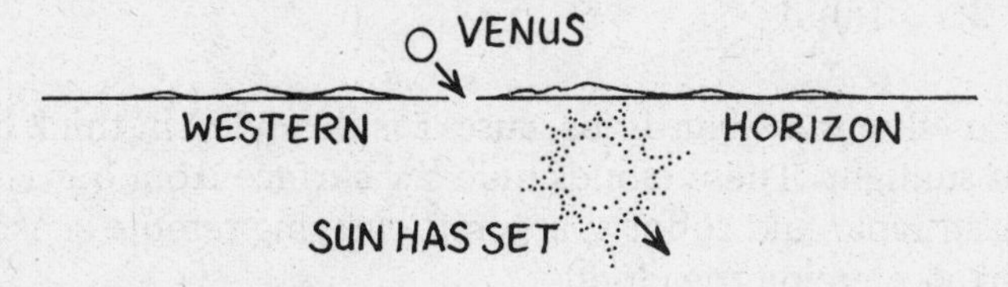

Figure 9.3. Venus as an evening star.

They are *morning stars* in the eastern sky just before sunrise near their western elongation. Then they lead the sun.

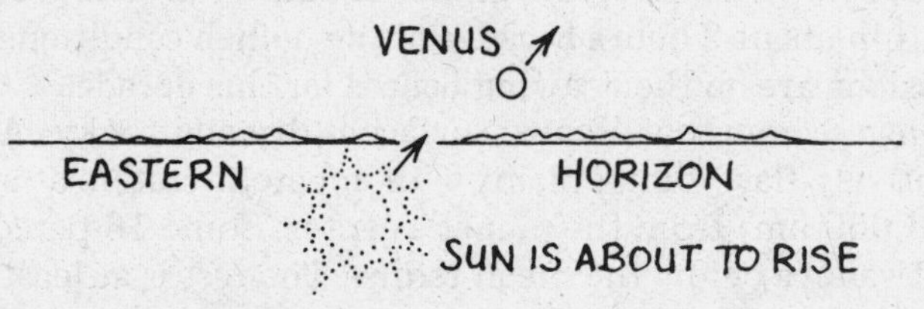

Figure 9.4. Venus as a morning star.

Normally Venus and Mercury pass above or below the sun at *conjunctions* (figure 8.7). About thirteen times in a century, Mercury, and less frequently Venus, *transit*, or pass directly in front of the sun at conjunction. Mercury will transit next on November 13, 1986, and Venus on June 8, 2004, and June 5, 2012. Observers will see a tiny dot moving across the bright face of the sun.

Venus is at *inferior conjunction* at intervals of 584 days. Then it comes closer to Earth (about 26 million miles away) than any other planet.

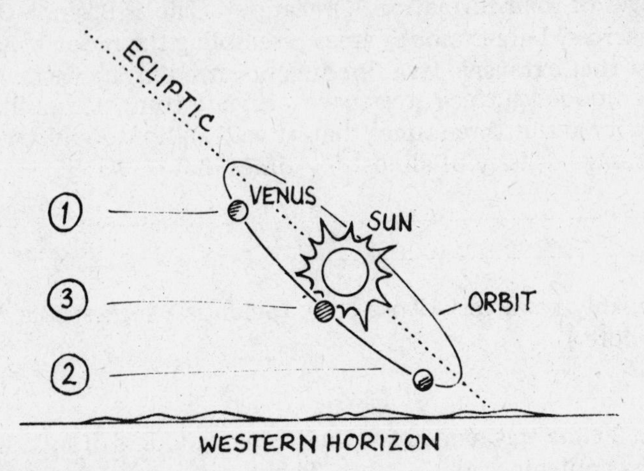

Figure 9.5. The orbit of Venus.

Referring to figure 9.5, which shows Venus in orbit, determine the planet's location when it is (a) an evening star ______; (b) a morning star ______; (c) at conjunction ______.

(a) 1; (b) 2; (c) 3

3. Venus shines brilliantly because it is shrouded in thick *clouds* that reflect a lot of sunlight. These clouds hide the surface from our view.

Earth-based radar and robot spacecraft carrying remote-sensing equipment have started piercing the clouds.

The latest American probe, *Pioneer Venus 1,* began orbiting Venus in December, 1978. It measured the terrain by radar and snapped ultraviolet and infrared pictures of the perpetual clouds. Soviet *Venera* landers took the first surface pictures ever in 1975; others gathered data in 1978 and 1982. All landers expired within about 2 hours because of the hellish conditions. More robot spacecraft missions are on the drawing boards for this decade.

Radar images show that Venus' surface is dry and rocky. About 60 percent is relatively flat, rolling plains with apparent craters at a mean radius 3,760 miles (6,050 km) from the planet's center. Some 16 percent is dry basins and rift valleys below the mean radius. The rest is at least a few thousand feet above the mean radius.

Figure 9.6. Venus as viewed by (*i*) Earth-based telescope, (*ii*) Pioneer Venus Orbiter, and (*iii*) Computer-generated global representation of radar topographical data.

Figure 9.7. First look at the surface of Venus from short-lived Soviet spacecraft.

Highlands that appear like continents tower above about 8 percent of the dry plains. *Terra Aphrodite* is the largest—half as large as Africa. Smaller *Ihstar Terra* is the size of the continental United States. Here *Maxwell Montes*, a mountain massif, soars highest of all—nearly 7 miles (11 km) above the mean radius. Apparently there are many fault zones and volcanoes.

The atmosphere is about 97 percent carbon dioxide, one to three percent nitrogen with traces of water vapor, helium, neon, argon, sulfur compounds, and oxygen. It circulates in large global motions. The temperature of the cloud tops is about -9° F (250° K). Apparently they are colored yellow by corrosive sulfuric acid. Cloud layers about 12 miles (19 km) thick (altogether) are about 40 miles (50–70 km) above the surface.

Venera landers found a very inhospitable world. Surface temperatures reach 900° F (755° K) because carbon dioxide and water vapor in the clouds trap the sun's radiation. This process is called the *greenhouse effect*. Atmospheric pressure is a crushing 1,330 pounds per square inch (over 90 atmospheres). There are many apparent lightning and thunder storms.

Venus is close to Earth in size, mass, density, and distance from the sun. However, you could not live there comfortably. Explain why not.

- - - - - - - - - - - - - - - - - - -

Venus (1) is much too hot, 900° F (755° K); (2) has a poisonous carbon dioxide atmosphere; and (3) has a crushing atmospheric pressure (over 90 atmospheres).

---

4. Our own planet *Earth* shines like a rare blue and white jewel in space. Third from the sun, it is the most important planet of all to us.

Figure 9.8. Earth from the moon.

The *total surface area* of our planet is almost 199 million square miles (5.10×10^8 km^2). More than 70 percent of our planet is covered by *water*, which is unique in the solar system.

The highest mountain on Earth is Mt. Everest in Asia, over 29,000 feet (8.8 km) above sea level. The deepest measured underwater spot is the Marianas Trench, more than 36,000 feet (11 km) below the Pacific Ocean's surface.

Earth's *mass* is about 6,000 sextillion kilograms (5.977×10^{24} kg). This mass provides the surface gravity we are used to.

Earth is somewhat pear-shaped. Its daily rotation around its axis has produced an *equatorial bulge* and *polar flattening.*

Refer to figure 9.9. How much longer in kilometers is the distance across the equator than the distance from North to South Pole?

About 43 kilometers.

Solution: equatorial diameter − polar diameter =
12,756.34 km − 12,713.80 km = 42.54 km.

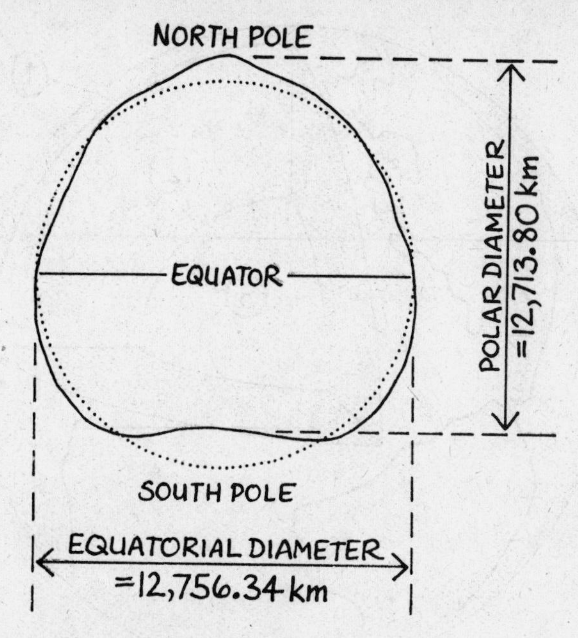

Figure 9.9. The pear-shaped quality of Earth.

5. Astronomers think that Earth was born about 4.6 billion years ago. It *formed* together with the other planets out of the same contracting cloud of gas and dust that formed the sun (frame 7.3).

Geologists picture the earth today in three layers. The solid surface layer is called the *crust.* The crust, an average of 22 miles (35 km) thick, contains the continents and oceans. It is composed mainly of lightweight rocks such as granite and basalt.

The next layer, about 1,800 miles (2,880 km) thick, is called the *mantle.* The mantle probably consists mostly of dense silicate rock that behaves somewhat like taffy—yielding under steady pressure but fracturing under impact.

The central layer, 2,170 miles (3,470 km) thick, is called the *core.* Here an outer liquid layer about 1,300 miles (2,080 km) thick surrounds a solid center. The core is probably made of dense iron and nickel at a temperature of about 6,400° K.

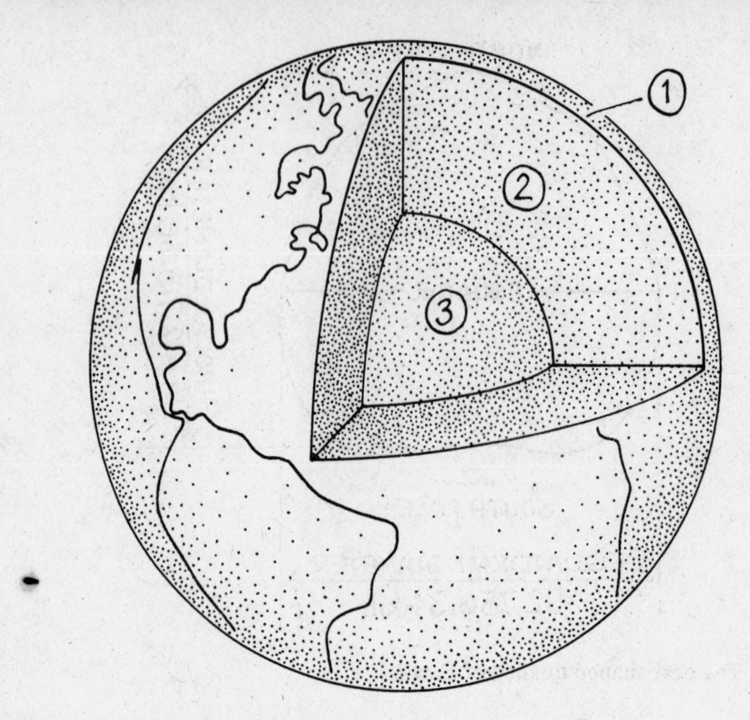

Figure 9.10. The structure of Earth

Referring to figure 9.10, identify the three principal layers of Earth and state the approximate thickness of each. ______________________

____

– – – – – – – – – – – – – – – – – – – –

(1) crust—an average of 22 miles (35 km); (2) mantle—about 1,800 miles (2,880 km); (3) core—about 2,170 miles (3,470 km)

6. The *surface* of our restless Earth is constantly *changing* because of erosion and geological activity. The oldest rocks discovered so far (in Greenland and Minnesota) are about 3.6 billion years old.

Substantial evidence indicates that about 200 million years ago all of the world's continents were joined in one huge *supercontinent* called Pangaea, which later broke up. According to the theory of *plate tectonics*, also called the *continental drift theory*, continents are embedded in rock slabs several thousand miles across called *plates.* The plates move slowly on the slightly yielding mantle beneath. As the plates move, the continents drift apart slowly, at about an inch (2 cm) per year. One inch a year adds up to over 3,000 miles in 200 million years.

Movement of the plates is also responsible for mountain building, earthquakes, and volcanic activity. These events occur at boundaries between the moving plates where they press on each other forcibly.

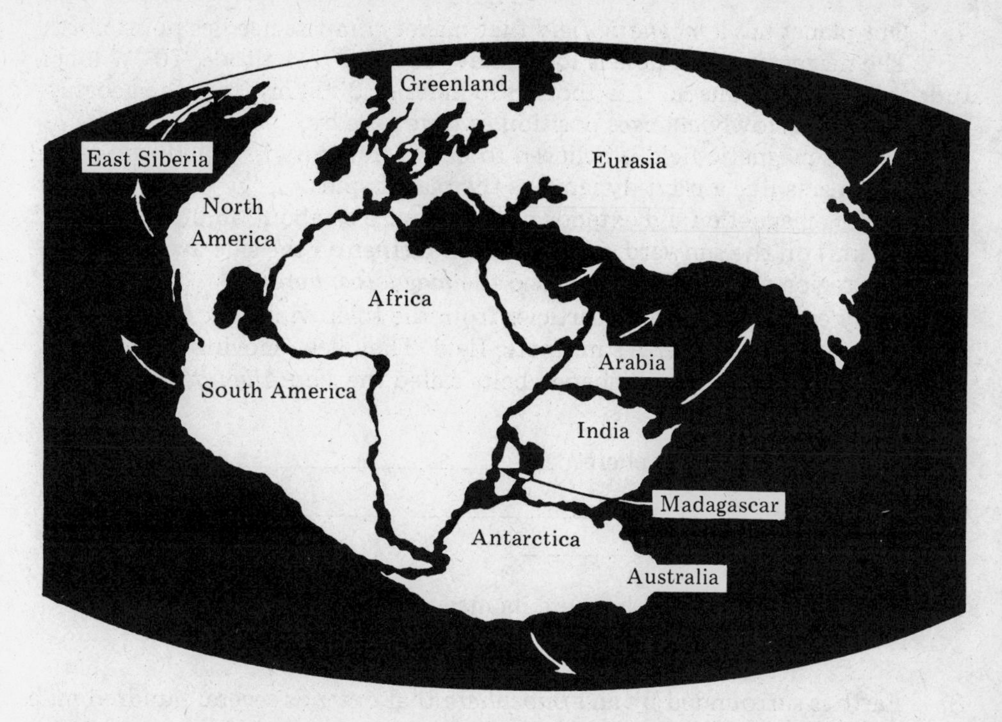

Figure 9.11. A map of Earth as it probably appeared about 200 million years ago.

A popular theory says that *magma convection currents* power the continental drift. Magma currents flow up through the mantle. Upon meeting cool rigid rocks, they flow horizontally. Friction drags the continent-bearing plates along. Finally the cooled magma sinks. Along mid-ocean ridges, magma pours through the crust creating new rocks perpetually.

Remarkable evidence that the continents drift is provided by the similar plant and animal fossils found along the coastlines of South America and West Africa. These coastlines seem to fit together, although they are now separated by almost 3,000 miles of Atlantic Ocean.

The ages of rocks from the bottom of the Atlantic Ocean have been measured recently. The oldest, found near the continental coastlines, are 150 million years old. If 4-billion-year-old rocks had been found on the bottom of the Atlantic Ocean, how would the continental drift theory have been affected? Explain. ______________________________

____

- - - - - - - - - - - - - - - - - -

Serious doubts would have been raised about the correctness of the theory, which says that the Atlantic Ocean, which is almost 3,000 miles wide between the continental coastlines, formed in the last 200 million years. It did not exist 4 billion years ago.

---

7. Our planet has a *magnetic field* that makes compass needles point north.

The magnetic north pole is located at about 79° N latitude, 70° W longitude in northeast Canada. It is about 830 miles (1,330 km) from the geographic north pole and slowly changes position as time goes by.

Earth's magnetic field is believed to be generated by its liquid iron-nickel core which acts like a giant dynamo as the planet spins.

Earth's magnetic field extends out into space to about 38,000 miles (60,000 km) on the sunward side and much farther on the side away from the sun. This region around Earth is called the *magnetosphere.*

Many energetic, charged particles from the solar wind that could be deadly are trapped by Earth's magnetic field. They keep moving around rapidly inside two doughnut-shaped belts called the *Van Allen belts* in the magnetosphere.

What is the magnetosphere? ____________________

____

- - - - - - - - - - - - - - - - - -

the region surrounding Earth where its magnetic field extends

8. Earth is surrounded by an *atmosphere* that extends several hundred miles out into space.

Earth's *primitive atmosphere* over 4 billion years ago was probably very different from air today. It may have contained noxious compounds of hydrogen, carbon, oxygen, and nitrogen such as carbon dioxide, ammonia, and methane, plus water vapor. The *free oxygen* we need for respiration probably came from photosynthesis in green plants.

Air today contains about 78 percent nitrogen, 21 percent oxygen, and 1 percent argon, carbon dioxide, and other gases. It also has variable amounts of water vapor, dust, carbon monoxide, chemical products of industry, and microorganisms.

Over half of this air is packed within the first 4 miles (6 km) above Earth's surface. The air thins out fast with increasing altitude. The *ozone layer* is about 12 to 50 kilometers above sea level. It contains the greatest concentration of ozone (three atoms of oxygen bound together). This layer shields Earth from deadly ultraviolet radiation from the sun.

The total mass of the entire atmosphere is about 5,000 trillion metric tons. Gravity keeps the atmosphere tied to Earth. At sea level all that air presses down with a force of 14.7 pounds per square inch, called 1 atmosphere of pressure. The *millibar* is another common unit of atmospheric pressure. At sea level, the air pressure on Earth is about 1,013 millibars.

What is the (a) composition and (b) pressure at sea level of the air that supports our lives on Earth? ____________________

____

- - - - - - - - - - - - - - - - - -

(a) about 78 percent nitrogen, 21 percent oxygen, and 1 percent carbon dioxide and other gases; variable amounts of water vapor and impurities.
(b) about 14.7 pounds per square inch, also called 1 atmosphere and 1,013 millibars.

9. Red *Mars* reminded the Romans of blood and fire, so they named it after their god of war. It has two small moons appropriately called Phobos (fear) and Deimos (terror), which can only be seen in powerful telescopes.

Superior planets like Mars look brightest when they are on the opposite side of Earth from the sun, a position called *opposition*. Then a fully lighted disk faces us. Mars is at opposition at intervals of 780 days on the average.

Superior planets are hardest to observe when they are on the opposite side of the sun from Earth, a position called *conjunction*.

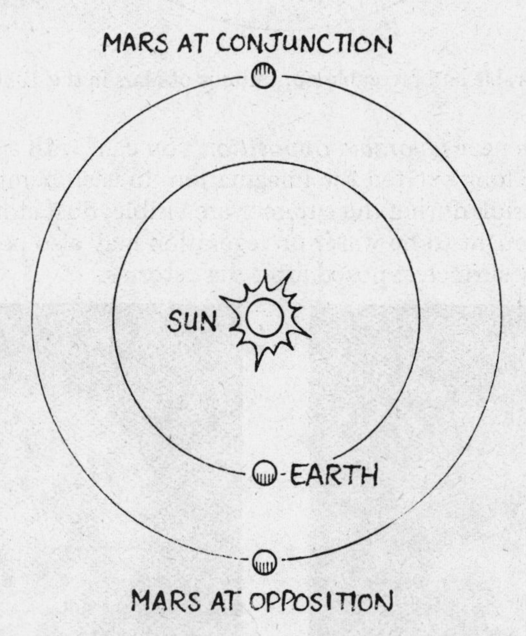

Figure 9.12. Mars at opposition and conjunction.

Mars comes closer to Earth at some oppositions than it does at others because of the eccentricity of its orbit (see Table 8.1). Close oppositions are called *favorable*, because the disk of Mars looks larger and observing is better. The most favorable oppositions occur when Mars is at perihelion. Then Mars is only about 35 million miles (56 million km) away from Earth. This happens in August at intervals of 15 to 17 years.

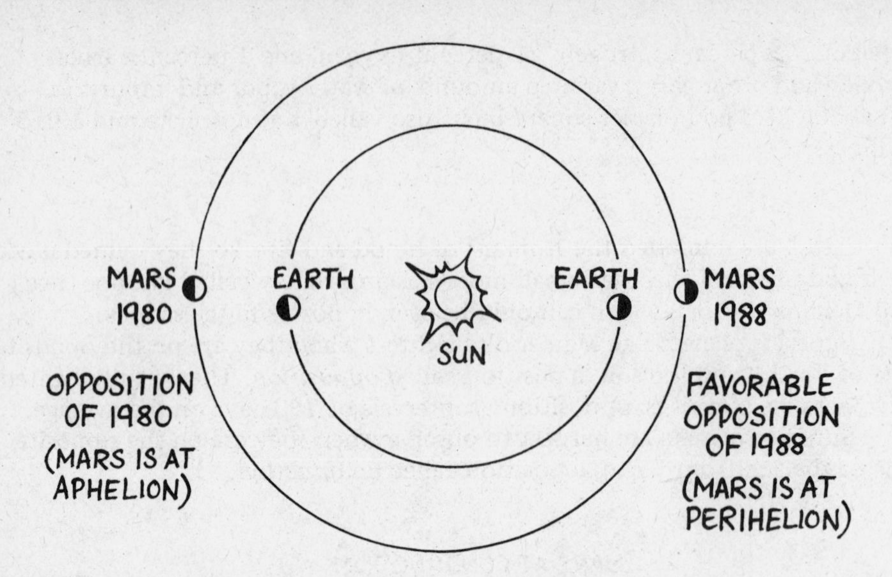

Figure 9.13. Unfavorable and favorable oppositions of Mars in the 1980's.

When Mars is near *favorable opposition* you can, with a telescope, see features that have long excited the imagination. In each hemisphere, white *polar caps* that shrink during the summer are visible; dust storms and dark areas formerly thought to be water or vegetation may also be seen. The dark areas are probably surface exposed after dust storms.

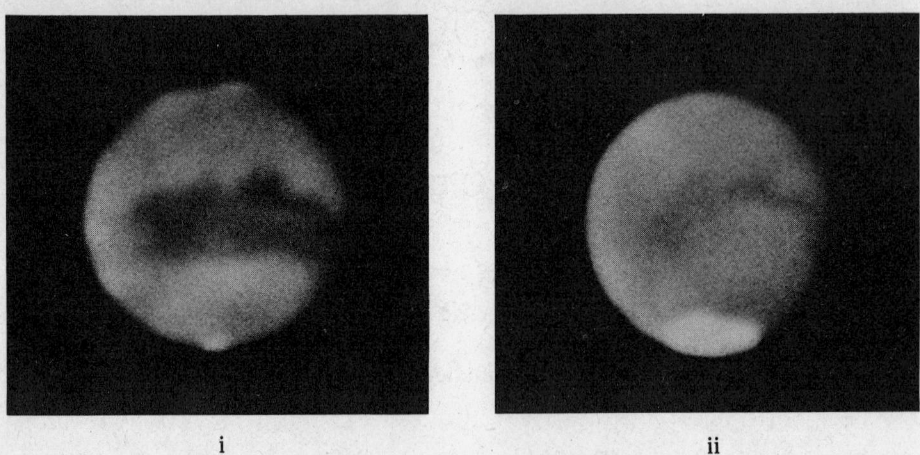

Figure 9.14. Mars through Earth-based telescope in (*i*) summer and (*ii*) winter.

Some observers have seen dark markings called *canals*, while others have not. "Canali" were reported first in 1877 by Italian astronomer G. V. Schiaparelli. This Italian word for "channels" was mistranslated into "canals" in English. American astronomer Percival Lowell caused great excitement at the beginning of the twentieth century when he said that intelligent Martians had built the canals. Space probe data show that canals are not present on Mars. These markings may actually be mountain ranges.

Why is Mars best observed at favorable oppositions?

It is closest to Earth then. (Both Mars and Earth travel in elliptical orbits around the sun, so the distance between them varies considerably—see figure 9.13.)

10. We took our first good look at the *surface* of another planet through the robot "eyes" of the Viking 1 lander which set down on Mars on July 20, 1976, seven years to the day after Apollo astronauts first walked on the moon.

Figure 9.15. First spectacular 300° panorama of the surface of Mars from Viking 1.

Scattered rocks, sand dunes, and distant low hills came into view when Viking 1 settled into the powdery dirt in a plain called Chryse Planitia. The site is located at 22.46°N latitude, 48.01°W longitude on Mars. Air temperature ranged from -122° F (-86° C) shortly after dawn to a high of -22°F (-30° C) at midafternoon on that Martian summer day. Air pressure was only about 7 or 8 millibars.

Two months later the Viking 2 lander set down in a plain called Utopia Planitia at 47.89° N, 225.86° W, about 4,600 miles (7,500 km) northwest of its predecessor. The site has rocks filled with bubbles that resemble lava from gaseous volcanoes on Earth.

The red dirt at the landing sites looks like iron-rich clay. Chemical weathering of the rocks and erosion seem to have occurred. Rocks are covered with fine-grained reddish material that is probably an iron oxide compound (like rust). The sky is colored pink in daytime by red dust that hangs in the atomosphere like smog. Sunsets are pale blue. In the winter temperatures drop below -190° F (-123° C) and a fine layer of frost appears.

No signs of large organisms were apparent at either site. There was no flowing water. Results of tests for microorganisms in the soil remain puzzling. (More on that in Chapter Twelve.)

Briefly describe the surface of Mars at the Viking lander sites. _______

____

— — — — — — — — — — — — — — — — — —

It looks like a red, dry, rock-strewn desert. The sky is pink. The temperature is cold.

11. The Viking 1 and 2 orbiters showed us a harsh, rugged, dry planet. Mars has huge volcanoes, and some may still be active. Giant Olympus Mons, or Mount Olympus, towers about 15 miles (24 km) above the surface. There are deep canyons. Valles Mariner, or Mariner Valley, is the largest—3,000 miles (5,000 km) long. Craters suggest that meteorites bombarded the planet billions of years ago. Some, such as 11 mile (18 km) wide Crater Yuty, look as if water and shattered rock poured out after a big impact.

The atmosphere is too thin to block the deadly ultraviolet rays from the sun that beat down on the planet. It is made up of about 95 percent carbon dioxide, with 2 to 3 percent nitrogen, 1 to 2 percent argon, 0.1 to 0.4 percent oxygen, with traces of water vapor and other gases.

Wild dust storms swirl out of the southern hemisphere in the summer. Often they rage over the whole planet. High speed winds blow light colored dust about, sculpting and exposing dark rock.

Mars does not have liquid surface water today. However, there is indirect evidence of ancient catastrophic flooding. Its deep winding channels resemble river beds with tributaries. They look as if they were carved long ago by great rivers. That water may be locked in the ice caps and permafrost under the surface today. Climate changes may have turned a running water environment into the cold, dry world Mars is today.

Viking 1 and 2 found direct evidence of water in solid and vapor form. The permanent ice cap at the north pole is made of frozen water. The ice cap at the south pole is also mostly frozen water, with some frozen carbon dioxide. The winter frost is apparently frozen water and dust. Further, there are occasional fog and filmy clouds.

Scientists think water is critical for life. Perhaps life formed on Mars in a past age that was warmer and wetter. Perhaps it still survives (more on this in Chapter Twelve).

List two pieces of evidence that indicate water once flowed on Mars.

____

— — — — — — — — — — — — — — — — — —

1. The deep, winding surface channels look as if they were carved by great rivers.
2. The permanent polar ice caps are almost entirely frozen water that may once have flowed on Mars.

Figure 9.16. Deep, winding channels and craters on Mars.

12. Phobos and Deimos are small irregular rock chunks only about 13 miles (21 km) and 9 miles (15 km) long, respectively. Phobos orbits Mars every 7

Figure 9.17. Phobos, the larger of Mars' two moons.

hours and 39 minutes, while Deimos completes a circuit in about 30 hours. Both have many craters and look fairly old. Phobos has striations and chains of small craters.

In 1727 Jonathan Swift wrote of two moons of Mars, long before they were actually discovered by American astronomer Asaph Hall (1829-1907) in 1877. Briefly describe the real moons of Mars.

– – – – – – – – – – – – – – – – – – –

small, irregular, cratered rock chunks

Figure 9.18. Collage of Jupiter and its four Galilean moons from Voyager 1.

13. Giant *Jupiter* was named for the mythological Roman king of the gods and ruler of the universe. The biggest planet of all, Jupiter at night outshines the stars and all the planets except Venus.

Jupiter has colorful, parallel, dark and light cloud bands and the *Great Red Spot*. Jupiter's four brightest moons (discovered by Galileo and collectively called the *Galilean moons*)—*Io*, *Europa*, *Ganymede*, and *Callisto*—change patterns nightly as they circle the planet. Astronomical publications (see Useful References) list moon positions, occultations, and transits.

Our best look at Jupiter came when the Voyager 1 and 2 space probes (frame 8.9) flew past the planet in 1979.

Jupiter's four brightest moons look like stars in a small telescope. What observations show they are really satellites of the planet? ______________

____

— — — — — — — — — — — — — — — — — — —

The moons are observed to change positions nightly as they circle the planet.

14. Jupiter is more massive than all the other planets and their moons combined. It barely missed being a star. If Jupiter were about eighty times more massive, nuclear fusion reactions could have started.

The planet seems to be a huge rapidly spinning liquid ball topped by a thick atmosphere, made of mostly hydrogen and helium. Apparently it has a relatively small iron silicate core. A thin ring of particles (from microscopic to a few meters in size) encircles Jupiter. It extends some 34,000 miles (57,000 km) out from the equatorial cloud tops.

Colorful changing cloud features and convoluted weather patterns circulate in the dynamic observable atmosphere. Superbolts of lightning flash. Complex patterns show in and between the moving dark-colored *belts* and lighter *zones*. Hydrogen, helium, and the detected traces of methane, ammonia, water vapor, and other gases in the atmosphere are all colorless. The bright red, orange, and brown clouds may be colored by different chemicals at various depths.

The famous *Great Red Spot* is a colossal atmospheric storm. It has been observed for over 300 years with varying size, brightness, and color. It rotates counterclockwise and also moves around the planet. The Great Red Spot dwarfs Earth, measuring about 9,000 miles (14,000 km) wide and up to 25,000 miles (40,000 km) long. Cooler than surrounding clouds, it towers up to 15 miles (24 km) above them.

Temperatures are as low as $-200°$ F at the cloud tops. The atmosphere extends down about 600 miles (1,000 km). Most likely the density of hydrogen increases steadily from the top inward until it changes to liquid hydrogen. Near the planet's center, the pressure and temperature could be high enough to compress hydrogen to an extraordinarily dense form called liquid metallic hydrogen. At the core temperatures may be 53,000° F (30,000° K), which would explain the observation that Jupiter radiates about twice as much heat as it receives from the sun.

The planet has a powerful magnetic field and a complex system of large, intense radiation belts somewhat like our Van Allen belts. These may account for some of Jupiter's observed radio emission.

The magnetic field is essentially dipolar but opposite Earth's in direction. (A compass needle on Jupiter would point south.) Electrical currents in the liquid hydrogen layer could be its source. At the cloud tops, Jupiter's magnetic field is 1.5 to 7 times more powerful than Earth's. Jupiter's enormous magnetosphere varies in size, possibly due to changes in the solar wind pressure. It may stretch sunward 9 million miles (15 million km) and outward nearly half a billion miles (690 million km), beyond Saturn's orbit at times.

Jupiter's atmosphere is especially interesting, because it may be similar to Earth's primitive one.

Of what is Jupiter's atmosphere made?____________________

____

– – – – – – – – – – – – – – – – – – – –

mostly hydrogen and helium, with traces of methane, ammonia, water vapor, and other gases

15. There are 16 known *moons* orbiting Jupiter (see Appendix 5 for details). Most are small. Voyager 1 and 2 focused on innermost Amalthea and the four large Galilean satellites.

Small *Amalthea* resembles a dark red football. Its elongated surface shows signs of meteorite impacts.

Colorful *Io* is the most geologically active world known beyond Earth. It has active volcanoes. Probably sulfur-rich materials spewed by volcanoes colors Io's array of red, orange, brown, blue, black, and white surface regions.

A gigantic cloud of charged particles, mostly ions of sulphur and oxygen, wobbles around Jupiter at Io's distance. The particles likely originate in Io's volcanic eruptions. Cloud particles may also travel along Jupiter's magnetic field lines into its north and south polar atmospheres causing the brilliant Jovian *auroras* observed by Voyager 1 and 2.

There is evidence of water ice on the surfaces of *Europa*, *Ganymede*, and *Callisto*. *Europa*, about the same size and density as our moon, is the brightest Galilean moon. Its smooth icy crust is crisscrossed by long lines that look like fractures. Ganymede and Callisto may be composed of up to 50 percent water, mixed with rocky material. *Ganymede* is the largest known moon in the solar system, 3,241 miles (5,216 km) in diameter. It has dark, probably ancient, areas with many craters like our moon's, and lighter younger terrain that is fractured and faulted suggesting past tectonic activity.

Callisto's surface looks oldest, with numerous impact craters. It may be over 4 billion years old. Features that look like the remains of very large basins may record collisions with large chunks of rock and metal.

(a) What is the largest known moon in the solar system?____________

(b) What is its diameter?____________________________

– – – – – – – – – – – – – – – – – – – –

(a) Ganymede (b) 3,241 miles (5,216 km)

Figure 9.19. Saturn and moons from Voyager 1.

16. *Saturn*, the most distant bright planet, was named for the Roman god of agriculture.

Like Jupiter, Saturn is a huge multilayered gas ball of mostly hydrogen and helium, with perhaps a relatively small iron-silicate core. Its dynamic atmosphere also has belts and zones. The details, color, and irregularities are less distinct because of a much thicker haze layer above the visible clouds. Saturn, too, is flattened at its poles by its rapid rotation. Like Jupiter, it gives off radio signals.

Highest speed winds, over 1,000 miles (1,600 km) per hour, occur at the equator and are much stronger than Jupiter's. Temperatures near the cloud tops are -305° F (near the center of the equatorial zone) to -294° F (86° K to 92° K). There are auroral emissions and lightening.

Dazzling *rings* surround Saturn. Although they stretch over 40,000 miles (65,000 km), the rings are only a few miles (km) thick. Stars can be seen through them. The rings probably consist of microscopic to meter-size particles that resemble icy snowballs or ice-covered rocks, orbiting Saturn. They shine by reflecting sunlight. We see the rings in different orientations as both Earth and Saturn orbit the sun.

The rings were named in order of their discovery. From the planet outward, they are known as D, C, B, A, F, G, and E. Hundreds of tiny ringlets comprise the A, B, and C rings which are seen in small telescopes. The F ring, first discovered by Pioneer 11 in 1979, is composed of three separate ringlets

Figure 9.20. The various aspects of Saturn from Earth.

which appear intertwined. Voyager 1 confirmed the existence of the D and E rings, and discovered the G ring in 1980. Long, spoke-like features in the B ring may be shiny fine particles raised by electrostatic forces.

Saturn's rings may be grains of material that never collected into a single moon. Or, they may be the remains of a moon that was torn apart. The *Roche limit* is the minimum distance a moon must be above its parent planet in order not to be torn apart by gravitational pulls. The rings of Saturn are closer to the planet than that minimum. That is, they are inside the Roche limit.

With a mass equal to about 95 Earths in a volume over 758 times Earth's, Saturn has the lowest average density of all the planets. It could float in water, if a big enough sea existed anywhere.

Saturn's magnetosphere is about one third as big as Jupiter's. It, too, varies in size when the solar wind intensity changes. It may extend sunward nearly a million miles (2 million km). The magnetic field drags along charged particles which circle Saturn as the planet rotates.

An enormous cloud of uncharged hydrogen atoms rings the planet between the orbits of Titan and Rhea. Inside this "doughnut," ultraviolet emitting particles orbit like miniscule moons.

Figure 9.21. Rings of Saturn from Voyager 1.

(a) What are Saturn's rings made of?____________________________

(b) Can you explain why they look solid in a small telescope? ____________

____

\- -

____

(a) probably icy microscopic to meter-size particles orbiting the planet in hundreds of ringlets

(b) The particles are so numerous and so *far away* from us. (Remember, distant galaxies look solid, too, although they are made of billions of separate stars.)

17. Saturn has at least 22 moons (see Appendix 5 for details). Others may be discovered as Voyager scientists study their voluminous encounter data.

The inner moons—*Mimas*, *Enceladus*, *Tethys*, *Dione*, and *Rhea*—are all smaller than our moon. They are apparently mainly water ice. Except for Enceladus, all are heavily cratered.

Titan is the only known moon with a substantial atmosphere. The atmosphere is mostly nitrogen, with hydrocarbons such as methane, ethane, acetylene, ethylene, and hydrogen cyanide. A dense haze hides Titan's surface. The moon is likely made of rock and ice.

The outer moons *Hyperion* and *Iapetus* seem to be mostly water ice. *Phoebe's* revolution is retrograde.

Newly discovered satellites seen in Voyager photographs are given temporary designations. Permanent names are finally assigned by the International Astronomical Union. At first little is known about these moons besides their orbits. Further analysis can reveal more. Watch for news about the composition of these moons and the nature of the gravitational interactions between the moons and rings.

Which is the only known moon in the solar system with a substantial atmosphere?____

- - - - - - - - - - - - - - - - - -

Titan

18. *Uranus* and *Neptune* look like giant twins. Both have thick hydrogen and methane cloud covers. Very little is known yet about these planets, because they are so far away.

Uranus was the first planet discovered using a telescope. William Herschel, in England, discovered it in 1781. Almost named for King George III, Uranus was finally named traditionally for the Greek god of the heavens.

Uranus, maximum magnitude +5.7, looks like a small disk (sometimes tinted green) in a telescope. You might spot it with your naked eye or binoculars, if you know exactly where to look (see Useful References).

The angle between the planet's axis and the pole of its orbit is a unique 98 degrees. So Uranus is practically lying on its side as it orbits the sun. Its rotation is retrograde.

At least nine rings orbit Uranus. Watch for more news about this planet and its satellites when Voyager 2 flies by it in 1986.

Neptune's discovery was a triumph for theoretical astronomy. Uranus did not follow the path Newton's law of gravity predicted it should. Astronomers John Adams in England and Urbain Leverrier in France calculated that its motion was being disturbed by another planet's gravity. They predicted where that unknown planet should be in the sky.

In 1846 astronomer Johann Galle at the Berlin Observatory pointed to the predicted spot and found Neptune. The planet was named for the Roman god of the sea. Neptune has two moons named Triton (a sea god) and Nereid (a sea nymph) (see Appendix 5 for details).

Figure 9.22. Artist's conception of the rings of Uranus.

Why was Neptune's discovery a triumph for theoretical astronomy?

— — — — — — — — — — — — — — — — — —

Theory predicted that Neptune (then unknown) must exist. The planet was discovered by looking at the spot in the sky predicted by theory.

Figure 9.23. Neptune and a satellite.

19. *Pluto*, farthest from the sun, is the most mysterious planet of all. It is very faint, maximum magnitude +14.

Irregularities in the orbits of Uranus and Neptune led to Pluto's discovery in 1930 by American astronomer Clyde Tombaugh. Pluto is probably a frozen, lifeless world, appropriately named for the Roman god who ruled over the dead beneath the Earth.

Pluto may shine as brightly as it does by reflecting sunlight from the observed methane and other ices that cover its surface. If so, Pluto may be even smaller than had been thought. Apparently it is not massive enough to be the proposed *Planet X* whose gravity disturbs Uranus and Neptune. Some astronomers still expect to discover Planet X.

Of all the planets, Pluto has the most *eccentric*, or elongated, orbit. At times this eccentric orbit takes Pluto inside Neptune's orbit where it is now. The planet is heading toward the sun and will reach perihelion in 1989. For the rest of this century, Pluto will be closer to the sun than Neptune.

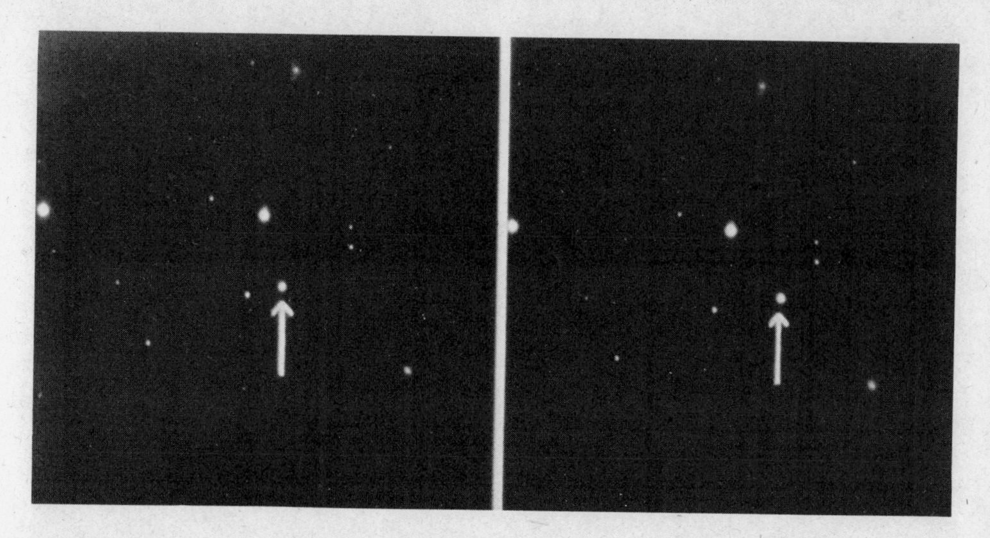

Figure 9.24. Two photographs of Pluto, showing motion of planet in 24 hours.

A bulge on Pluto's image seen in photographs by American astronomer James W. Christy in 1978 is apparently a moon. The moon is named *Charon* for the mythological boatman who ferried souls of the dead to Hades. Charon is 10,000 miles (17,000 km) above Pluto. It circles Pluto in 6 days, 9 hours.

Pluto is so far out in the solar system that it has not completed even one trip around the sun since its discovery. Look at figure 9.24. Can you explain how astronomers can verify that Pluto is not a star?

- - - - - - - - - - - - - - - - - - - -

Photographs of Pluto taken at different times show how it "wanders" relative to the background stars.

SELF-TEST

This self-test is designed to show you whether or not you have mastered the material in Chapter Nine. Answer each question to the best of your ability. Correct answers and review instructions are given at the end of the test.

1. Match each planet to a famous feature visible in a small telescope.

___ (a) phases	1. Mars
___ (b) polar ice caps	2. Jupiter
___ (c) Great Red Spot	3. Saturn
___ (d) rings	4. Venus

2. Match common features to correct planet pairs.

___ (a) alternate, parallel, dark and light cloud bands	1. Mercury and Venus
___ (b) many craters and mountains	2. Jupiter and Saturn
___ (c) thick hydrogen and methane cloud covers	3. Uranus and Neptune

3. List three reasons why Venus would be a very unpleasant planet to visit.

___

4. The diagram shows Venus, Earth, and Mars in orbit around the sun. What letter in the diagram indicates the following points?

___ (1) Venus is an evening star.

___ (2) Venus is in a new phase.

___ (3) Mars is at opposition.

___ (4) Mars is not visible in our nighttime sky.

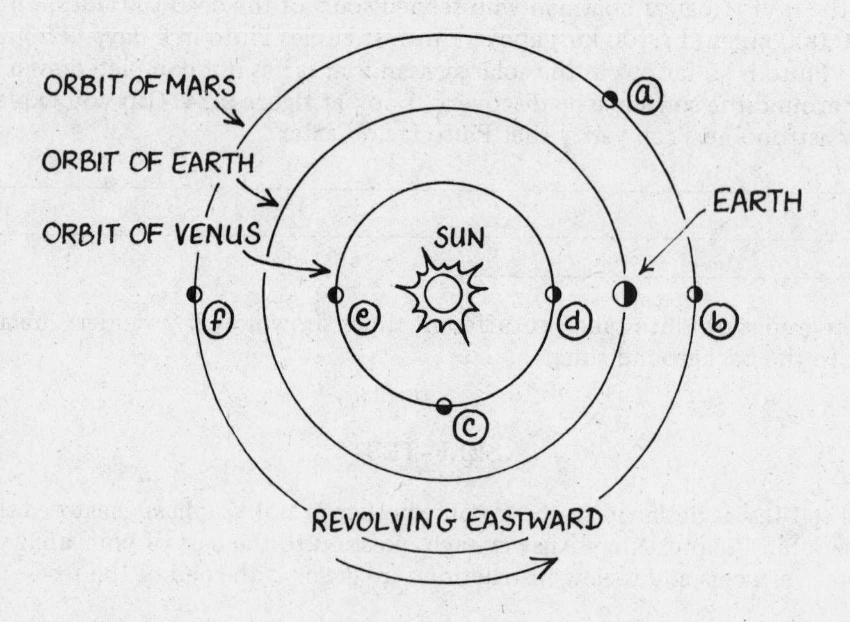

Figure 9.25. Venus, Earth, and Mars in orbit around the Sun.

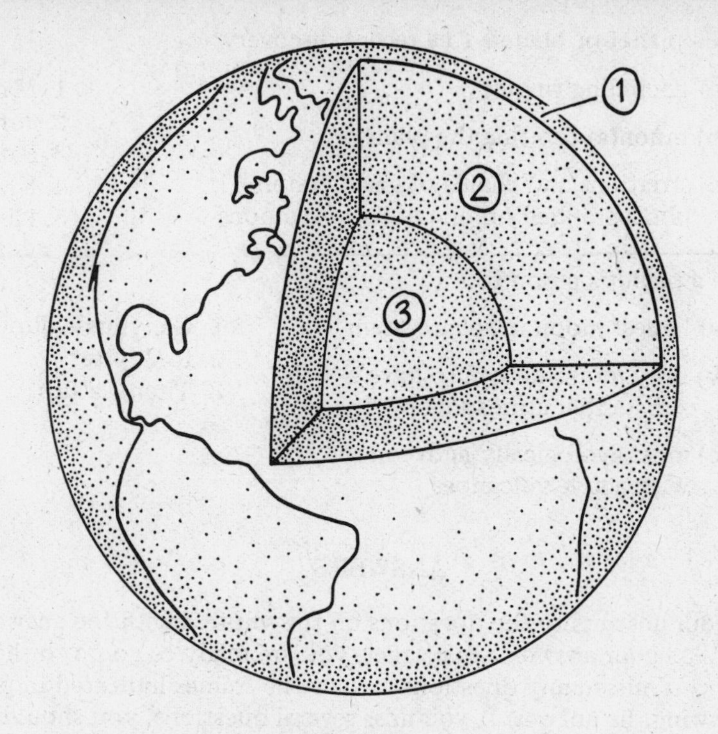

Figure 9.26. The structure of Earth.

5. Label the three principal layers of the earth.

   (1) ______________ (2) ______________ (3) ______________

6. Give three observations that support the theory of plate tectonics (continental drift). ______________

   ____

7. Describe the scene, atmosphere, and temperatures at the Viking 1 landing site on Mars. ______________

   ____

8. State two observations that indicate that water may have flowed on Mars long ago. ______________

   ____

9. List the most abundant gases in the atmospheres of

   (a) Earth ______________

   (b) Mars ______________

   (c) Jupiter ______________

   (d) Titan ______________

10. Match a planet or planets to a recent discovery:

    ___ (a) encircling ring(s)

    ___ (b) moon(s) orbiting the planet

    ___ (c) striations and chains of small craters photographed on one of its two moons

    1. Mars
    2. Jupiter
    3. Saturn
    4. Uranus
    5. Pluto

11. Match a planet's moon to:

    ___ (a) largest moon in solar system

    ___ (b) only moon known to have a substantial atmosphere

    ___ (c) most geologically active moon, with active volcanoes

    1. Ganymede/Jupiter
    2. Io/Jupiter
    3. Titan/Saturn

ANSWERS

Compare your answers to the questions on the self-test with the answers given below. If all of your answers are correct, you are ready to go on to the next chapter. If you missed any questions, review the frames indicated in parentheses following the answer. If you miss several questions, you should probably reread the entire chapter carefully.

1. (a) 4; (b) 1; (c) 2; (d) 3 (frames 2, 9, 13, 14, 16)
2. (a) 2; (b) 1; (c) 3 (frames 1, 3, 13, 14, 16)
3. poisonous carbon dioxide atmosphere; much too hot (up to 900° F); and a crushing atmospheric pressure over 90 atmospheres) (frame 3)
4. (1) c; (2) d; (3) b; (4) f (frames 2, 9)
5. (1) crust; (2) mantle; (3) core (frame 5)
6. (1) Similar plant and animal fossils are found along coastlines of South America and West Africa. (2) These coastlines seem to fit together. (3) No rocks from bottom of Atlantic Ocean near the coastlines are older than about 150 million years. (frame 6)
7. The surface looks like a red, dry, rock-strewn desert. The sky is pink. The temperature is cold. (frame 10)
8. (1) deep, winding channels that look as if they were carved by great rivers; (2) water frozen in the polar ice caps (frame 11)
9. (a) nitrogen (about 78 percent) and oxygen (about 21 percent); (b) carbon dioxide; (c) hydrogen and helium; (d) nitrogen (frames 8, 11, 14, 17)
10. (a) 2, 3, 4; (b) 2, 3, 5; (c) 1 (frames 12, 15, 17, 19)
11. (a) 1; (b) 3; (c) 2 (frames 15, 17)

CHAPTER TEN

The Moon

The stars about the lovely moon hide their shining forms when it lights up the earth at its fullest.

Sappho (c. 612 B.C.)
Fragment 4

1. Poets have always been enchanted by the beautiful full moon. At *magnitude −12.5*, it is almost 25,000 times more brilliant than first magnitude stars.

Figure 10.1. The full moon.

People once believed that the brilliant moon influenced personal behavior directly. They practiced special rituals at full moon. Words such as "moon-struck" and "lunacy" originally referred to a madness that changed with the phases of the moon.

Today we know more about the moon than about any other neighbor in space. It is the closest sky object of all, located an *average* of 240,000 miles (384,404.377 km) from Earth. Six *Apollo moon missions* (1969-1972) sent man there with scientific experiments and returned 843 pounds (382 kg) of moon rock for laboratory study.

The moon's *albedo*, or the fraction of incident sunlight that the moon reflects back out into space, is only 0.07. Most of the sunlight that shines onto the airless moon's surface is absorbed.

Why do you think the full moon is the brightest light in the night sky?

____

— — — — — — — — — — — — — — — — —

Because it is so much closer to Earth than all other sky objects.

2. If you look at the moon regularly, you will observe its *two apparent motions* in the sky in addition to its phases (frame 8.4).

You will see the moon rise in the east, move *westward across the sky*, and set every day, because the earth rotates daily.

You will also observe that the moon changes its location with respect to the stars about *13° to the east* every day, because the moon moves with respect to the sun daily as the Earth-moon system revolves around the sun every year, as shown in figure 10.2.

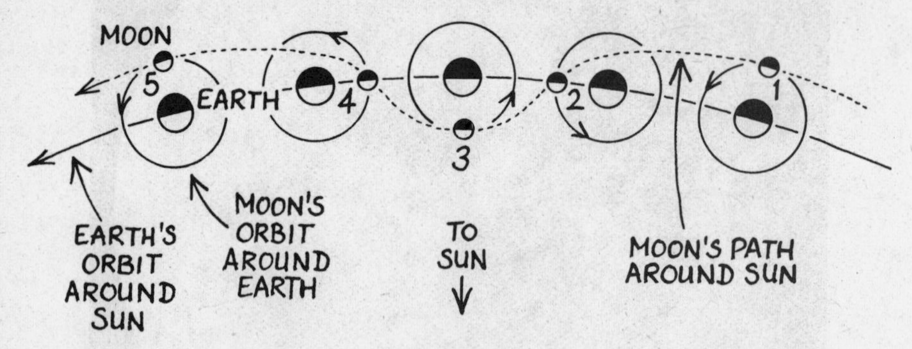

Figure 10.2. The Earth-moon system's revolution around the sun. The waviness of the moon's orbit is greatly exaggerated for clarity.

Explain why the moon rises an average of about 50 minutes later each day than it did the day before. ______________________________

____

— — — — — — — — — — — — — — — — —

At moonrise the moon is located in a particular constellation. About 24 hours later, when Earth has turned completely around, those same stars rise again, but meanwhile the moon has moved about 13° eastward with respect to the stars and so does not rise until later.

3. Earth's gravity has locked the moon into *synchronous rotation.* The moon *rotates* on its axis every $27\frac{1}{3}$ days, the same amount of time it takes to travel around Earth. That makes the same side of the moon face Earth at all times.

That means you observe the same features of the "man in the moon" all month long, but never see the back of his head.

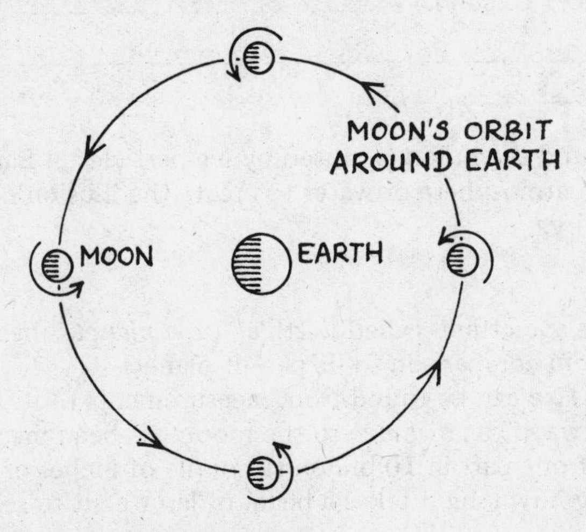

Figure 10.3. The moon's synchronous rotation.

The moon's *rotation period* and its *revolution period* are probably not equal only by coincidence but are most likely equal because of the Earth's gravitational pull on the moon.

Why were people able to see only one side of the moon before spacecraft flew to the back side? ______________________________

____

– – – – – – – – – – – – – – – – – – – –

The moon's rotation period is just equal to its period of revolution around Earth, so the same side of the moon always faces Earth.

4. You may have wondered what causes some other changes in the moon's appearance.

The lunar *halo*, or *ring around the moon*, is really not near the moon at all. Ice crystals high up in Earth's atmosphere *refract*, or bend, moonlight as it passes through, creating the halo effect.

When the moon is near the horizon, it may look *red.* From that position, moonlight travels a longer path through the atmosphere to our eyes than when the moon is overhead. Moonlight (reflected sunlight) consists of all visible colors. Short (blue) moonlight rays are *scattered* out and the long (red) rays, which penetrate the atmosphere more readily, color the moon red.

The *harvest moon* is the full moon nearest the time of the autumnal equinox. It rises earlier in the evening than usual, lighting up the sky to give extra hours for fall harvesting. Harvest moon occurs when the angle between the ecliptic (frame 8.2) and the horizon is near minimum.

Do you think the Apollo astronauts saw a "ring around the earth" from the moon's surface? Explain. ______________________________

____

- - - - - - - - - - - - - - - - - -

No. The ring around the moon is caused by ice particles in Earth's atmosphere. The moon has no atmosphere or water to create the illusion of a ring around the earth in the sky.

5. The moon is sometimes called Earth's "*twin planet,*" because our satellite is unusually large in comparison to its parent planet.

The moon's *size* can be found from measurements of its angular diameter and its distance away. The *distance* to the moon has been measured to a fantastic accuracy of one part in 10 billion (a couple of inches or a few centimeters) by timing how long it takes a beam of laser light to reach reflectors there and return.

The *diameter of the moon* is found to be about 2,170 miles (3,476 km). The *diameter of Earth* is almost 8,000 miles (12,756 km). Compare the size of the moon to that of Earth. ______________________________

- - - - - - - - - - - - - - - - - -

The diameter of the moon is about $\frac{1}{4}$ that of Earth.

Solution: $$\frac{\text{diameter of moon (approximately 2,000 mi.)}}{\text{diameter of Earth (approximately 8,000 mi.)}} = \frac{1}{4}$$

6. The moon's *mass*, measured from the accelerations the moon produces on spacecraft, is 7.35×10^{25} grams, or about $\frac{1}{81}$ that of Earth.

The moon's average density is 3.34 grams per cubic centimeter, or roughly $\frac{3}{5}$ that of Earth.

The moon's *surface gravity* is only about $\frac{1}{6}$ that of Earth because of its small mass and size. That means a 180-pound astronaut would weigh only 30 pounds ($\frac{1}{6}$ of his Earth weight) on the moon's surface.

Suggest a reason why the moon's average density is less than Earth's.

Probably they have different chemical compositions. (Moon rocks collected so far are made of the same chemical elements as Earth rocks, but the proportions are different.)

7. Review the properties of the moon you have learned so far by completing for yourself the following handy reference table (Table 10.1). Review frames 8.1 and 8.6, if necessary.

Table 10.1. Properties of the moon

Quantity	Value
(a) average distance from Earth	
(b) diameter	
(c) sidereal period (fixed stars)	
(d) synodic period (phases)	
(e) rotation period	
(f) mass	
(g) average density	
(h) surface gravity	
(i) albedo	
(j) apparent magnitude of full moon	
(k) average velocity in orbit	

(a) 240,000 miles (384,404.377 km); (b) 2,170 miles (3,476 km), or $\frac{1}{4}$ that of Earth; (c) $27\frac{1}{3}$ days ($27^d 7^h 43^m 11^s$); (d) $29\frac{1}{2}$ days ($29^d 12^h 44^m 2^s.9$); (e) $27\frac{1}{3}$ days; (f) 7.35×10^{25} grams, or $\frac{1}{81}$ that of Earth; (g) 3.34 grams per cubic centimeter; (h) $\frac{1}{6}$ that of Earth; (i) 0.07; (j) −12.5; (k) 2,295 miles per hour (1.02 km/sec)

8. The moon has long been a favorite target for *small-telescope observers*, because it is close enough to see in great detail.

When Galileo first pointed his telescope moonward, he mistakenly thought the large dark areas he saw were oceans. He called them *maria* (singular: *mare*), meaning seas.

The Apollo moon missions found no water at all on the moon. The roughly circular maria are actually dry lava beds.

The maria contain *basalt*, a type of rock produced by the cooling of molten lava from volcanoes. Mare Imbrium, the Sea of Showers, the largest mare on the moon's visible side, is about 1,700 miles (1,100 km) across.

The light-colored parts of the moon are called *highlands.* They are higher, more rugged, older regions than the maria.

What are the maria that form the features of the "man in the moon?"

- - - - - - - - - - - - - - - - - - -

solidified lava beds

9. You can see that the moon is pitted with *craters*, or holes, in the surface (see figure 10.5, for example).

Typical craters are circular, ranging in size from tiny pits to huge circular basins hundreds of feet across with walls ranging up to 10,000 feet (over 3,000 m) in height. They are customarily named after famous scientists and philosophers such as Copernicus and Plato.

Most craters were probably blasted out by high-speed meteorites crashing onto the moon. The largest craters, such as Clavius, nearly 150 miles (240 km) across, are called *walled plains.* The smallest are known as *craterlets.*

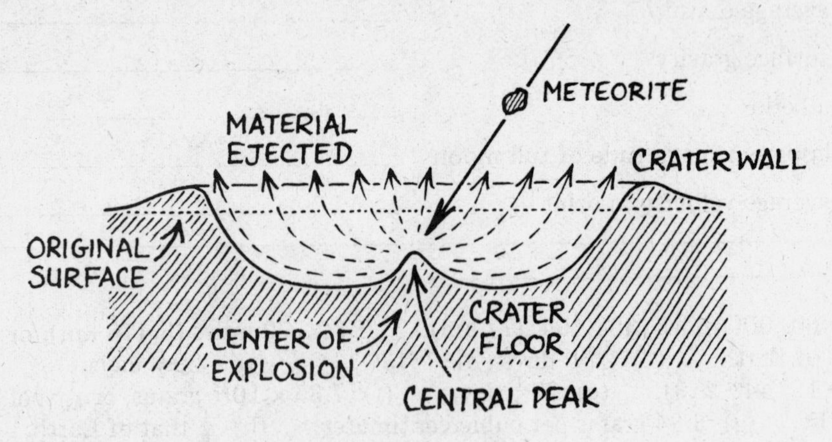

Figure 10.4. Formation of a crater by meteorite impact.

The best time to observe craters with your telescope or binoculars is when they are near the line called the *terminator*, which separates day from night on the moon.

Spacecraft photographs show that the far side of the moon does not have large maria, which are so conspicuous on the near side. The cause of the observed differences between the front and back sides of the moon is not yet understood.

Figure 10.5. The back side of the moon.

What probably produced most craters on the moon?

---

meteorites crashing into the surface

10. When Neil Armstrong first set foot on the moon for "all mankind," on July 20, 1969, he entered a strange desolate world.

The entire surface of the moon appears to be covered by basalt, or cooled lava. The highland rocks are older than those in the maria, however. No water, fossils, or organisms of any kind have been found in all the moon rocks and soil from the Apollo and Soviet Luna (robot) flights. This lack of evidence of life suggests that the moon is and has always been lifeless.

No blue sky, white clouds, or weather of any kind appear above the moon because the moon has no appreciable atmosphere. Ghostly silence prevails in the absence of air to carry sounds.

Days and *nights* are long—fourteen earth-days each. The *surface temperature* ranges from about 250° F when the sun is at its highest point to about -280° F at night.

Figure 10.6. Astronaut on the moon.

Why would it be useful to place a large optical telescope on the moon?

____

____

In the absence of air or weather on the moon, the "seeing" (frame 2.21) would always be good.

11. The major agents of *erosion*, or wearing away, of the moon's surface, are *micrometeorites*, tiny grains of rock and metal, that crash into the moon at speeds of up to 70,000 miles an hour.

Typical micrometerorites range in size from one ten-thousandth to one-thousandth of an inch. Large meteorites collide with the moon, too.

Micrometeorites are about 10,000 times less effective in changing the moon's surface than are air and water on Earth. They remove only about 1 millimeter of lunar surface in a million years.

Figure 10.7. Neil Armstrong's footprint on the moon.

Explain why Neil Armstrong's first footprint on the moon will probably look the same millions of years from now as it did in 1969 (see figure 10.7). ______________________________

- - - - - - - - - - - - - - - - - -

Erosion on the moon is due primarily to micrometeorite bombardment and happens much more slowly than erosion by air and water on Earth.

12. *Mountains* on the moon are named after the great ranges on Earth, such as the Alps.

Lunar mountains are really different from ours in both chemical composition and appearance. They were probably produced and shaped by different forces.

The highest rugged mountain peaks on the moon tower up to 29,000 feet (over 8,000 m), as does Mount Everest, Earth's highest mountain.

What two major factors that constantly change the shape of Earth's mountains do not shape moon mountains? Explain. ____________

____

- - - - - - - - - - - - - - - - - -

water and atmosphere. No mountain streams pour down these ranges. No atmospheric storms ever rage to wear away the surface.

13. Lunar scientists have reconstructed this *life story* of the moon from Apollo moon-flight data.

The *oldest* moon rocks, collected in the highlands, are about 4.3 billion years old. A few tiny green rock fragments are about 4.6 billion years old. The youngest, from the maria, were formed about 3.1 billion years ago. Moon rocks are richer in some minerals and poorer in others than Earth rocks. This casts doubt on the theory that the moon is a part of Earth that was torn off.

The manner and place of birth of the moon remain a mystery. It probably formed by the accretion of many smaller particles in the solar nebula.

During its first billion years of life, the young moon was heavily bombarded by meteorites of all sizes. They produced great craters and melted the surface, now the *lunar crust.*

When the moon was about 1 billion years old, the interior was *heated* so much by *radioactive elements* that *volcanoes* poured huge floods of hot basaltic lava over the surface and into the craters. This molten lava solidified, forming the maria.

About 3 billion years ago the moon cooled off significantly, and volcanic activity largely stopped. Except for minor lava flows and a relatively small number of large impact craters like the young (about a billion years old) Copernicus, the moon has not changed since. *Seismometers* (instruments that detect earthquakes) left on the moon by Apollo astronauts have detected a very low level of moonquakes.

How does the story of the moon's geological activity differ from Earth's?

____

____

- - - - - - - - - - - - - - - - - - -

The moon became essentially dead geologically after the first 2 billion years of its life in contrast to Earth, which is still very much alive with volcanism, mountain building, and drifting continents.

14. Geologists draw their current picture of the *moon's interior* from Apollo flight data.

Mascons (*mass concentrations*, or clumps of mass) were detected submerged in the circular maria. The existence of mascons plus the absence of major moonquakes suggest that the moon has a cold, thick, rigid outer layer, or *crust*, about 60 kilometers thick.

Beneath the crust, extending down about 1,000 kilometers, is the *mantle*. The physical characteristics of the *core*, extending the last 700 kilometers to the center are still unknown. The core is probably about 1,500° K and partly molten.

The moon does not have a magnetic field today, but old lunar rocks indicate that it once did.

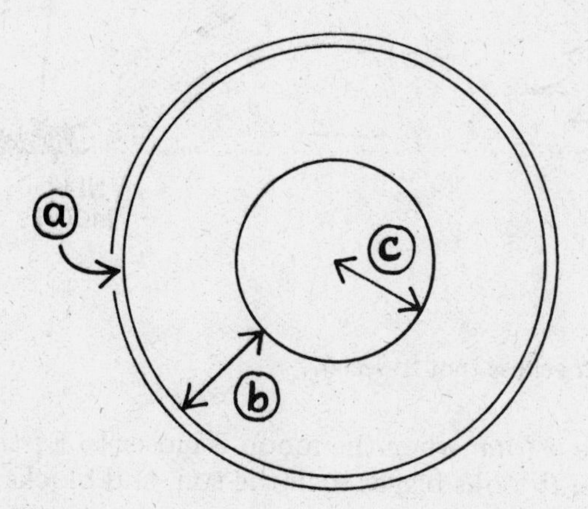

Figure 10.8. The moon's interior.

Referring to figure 10.8, identify the crust, mantle, and core, and indicate the approximate depth of each layer.

(a) ________________; (b) ________________; (c) ________________

- - - - - - - - - - - - - - - - - -

(a) crust (60 km); (b) mantle (1,000 km); (c) core (700 km)

15. More research on the moon is expected in the future. Much of the lunar material brought back from Apollo and Luna missions has not been analyzed yet. Apollo instruments left on the surface sent back data until they were finally turned off by NASA in 1977. One day scientists may return to the surface for further exploration. Write a brief summary paragraph describing what we have learned about the surface of the moon so far.

____

____

____

____

- - - - - - - - - - - - - - - - - -

Your paragraph should describe the maria, craters, mountain ranges, absence of air and water, length of day and night, and surface temperatures.

16. A *solar eclipse* occurs when the earth, new moon, and sun are directly in line.

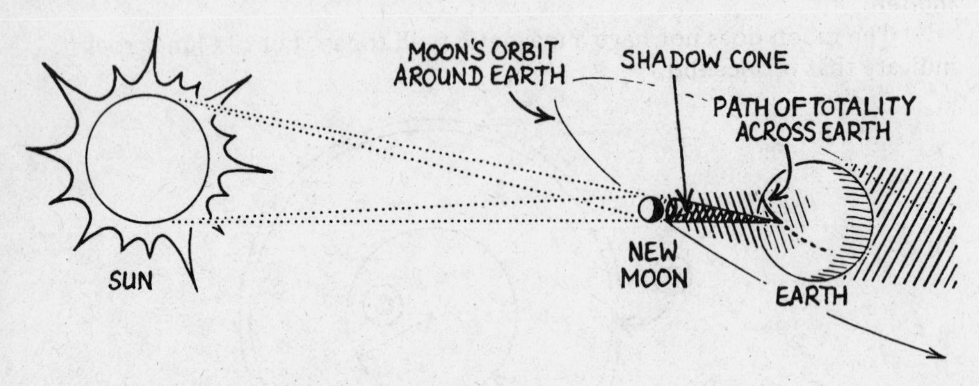

Figure 10.9. Solar eclipse (not to scale).

The eclipse is *total* when the moon is closer to Earth than the length of its shadow cone. It looks bigger than the sun, and blocks the sun's disk from view.

Annular Eclipse

Figure 10.10. Annular solar eclipse.

An *annular eclipse* occurs when the moon is farther from Earth than the length of its shadow cone. It looks smaller than the sun and blocks all but an outer ring of sunlight from view.

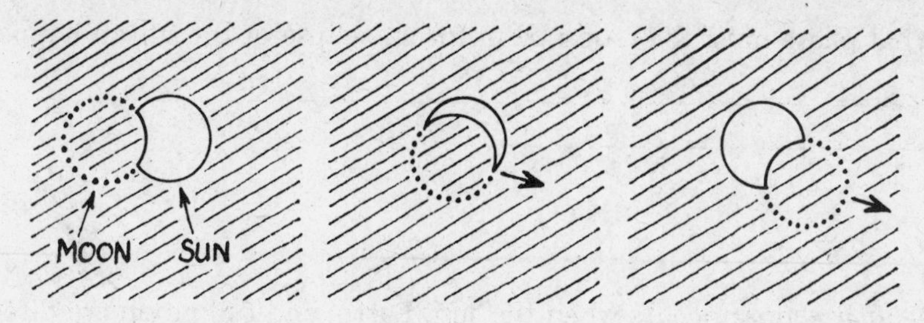

Figure 10.11. Partial solar eclipse.

A *partial eclipse*, the most frequent kind, occurs when the moon is not quite close enough to the sun-Earth line to block all of the sun from view.

To observe a total eclipse of the sun is thrilling. When the moon passes in front of the bright sun, an unnatural darkness spreads across the sky, the temperature drops, and stars and planets shine in the daytime sky.

Totality lasts only a few minutes and can be seen only from a narrow *eclipse path* on Earth. Your chances of observing a total solar eclipse from your hometown are very small, since the occurrence at any one location on Earth averages only about one in 360 years.

People once connected the sun's disappearance with terrifying events. Today professional and amateur astronomers eagerly go around the world to observe a total solar eclipse and gather astronomical data. You might consider joining an eclipse expedition yourself to view this spectacular natural event. The most important coming solar eclipses are listed in Table 10.2.

Table 10.2. Total solar eclipses

Date	Duration of Totality (minutes)	Where Visible
1983 June 11	5.4	Indonesia
1984 Nov. 22	2.1	Indonesia, South America
1987 March 29	0.3	Central Africa
1988 March 18	4.0	Philippines; Indonesia
1990 July 22	2.6	Finland, Arctic regions
1991 July 11	7.1	Hawaii, Central America, Brazil
1992 June 30	5.4	South Atlantic
1994 Nov. 3	4.6	South America
1995 Oct. 24	2.4	South Asia
1997 March 9	2.8	Siberia, Arctic
1998 Feb. 26	4.4	Central America
1999 Aug. 11	2.6	Central Europe, Central Asia

What phase must the moon be in for an eclipse of the sun to happen?

---

new

17. A *lunar eclipse* occurs when the sun, Earth, and full moon are directly in line.

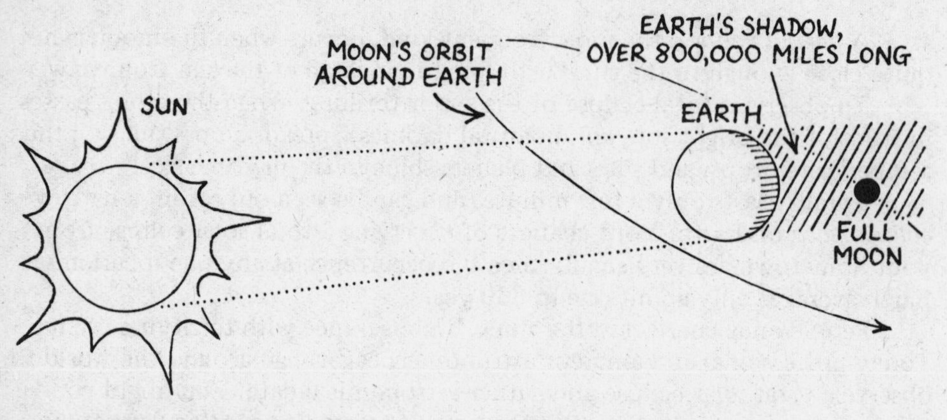

Figure 10.12. Lunar eclipse (not to scale).

The moon darkens when it enters Earth's shadow, but still gets some sunlight that is refracted (bent) around Earth by our atmosphere. Clouds, dust, and pollution affect the color (usually making it dull red) and brightness of the moon's appearance.

Your chances of seeing a total lunar eclipse are much greater than your chances of seeing a total solar eclipse. Lunar eclipses, when they occur, can be seen at any place on Earth where the moon is shining, and they last over an hour. Check current annual publications (Useful References) for future lunar eclipses.

More than 2,000 years ago, the Greeks observed that during a lunar eclipse, Earth's shadow appears circular on the moon. Aristotle (384-322 B.C.) cited this evidence in his writings to support the theory that the Earth is a sphere rather than flat. The first fairly accurate measurement of the diameter of the earth was made by a Greek astronomer, Eratosthenes (c. 276-195 B.C.).

Although astronomers do not attach major scientific importance to them, lunar eclipses are fun to watch.

What phase must the moon be in for a lunar eclipse to occur? ________

---

full

18. The *maximum number* of solar and lunar eclipses that can occur in one year is seven.

Eclipses do not occur every time we have a new or full moon, as you might expect. The *moon's orbit is tilted* 5.2 degrees to the plane of Earth's orbit. Most months the moon is above or below the sun-Earth line at new and full phase, so no eclipse can occur.

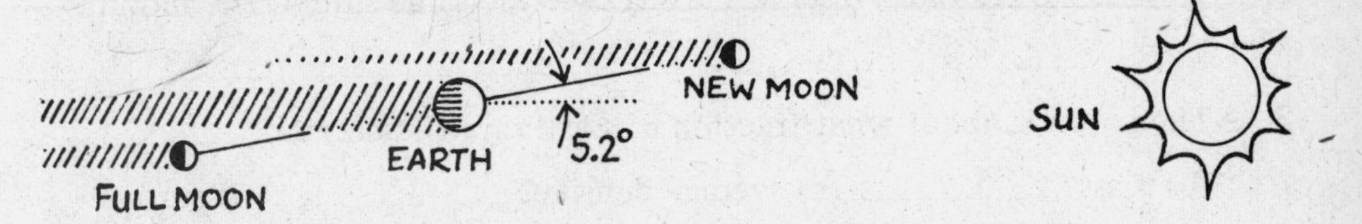

Figure 10.13. Unfavorable conditions for an eclipse.

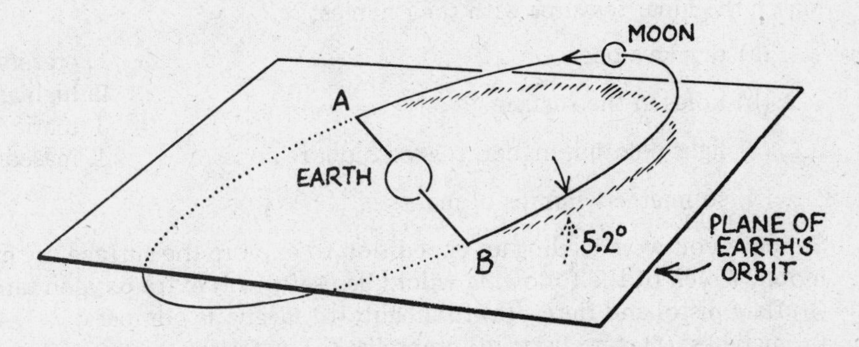

Figure 10.14. The tilt of the plane of the moon's orbit to the plane of Earth's orbit.

Examine figure 10.14. Explain why an eclipse can occur only when the moon is at point A or point B.

Then the sun, Earth, and moon are directly in line.

19. *Occultation* means the eclipse of one sky object by another, as when a star or planet is briefly hidden from view when the moon passes in front of it. This phenomenon is important in the observation of radio sources and in pinpointing their locations in the sky.

Jupiter is over forty times bigger than the moon. How is it possible for the moon to occult, or hide, Jupiter? ____________________

Jupiter is much farther away than the moon, so it appears much smaller.

SELF-TEST

This self-test is designed to show you whether or not you have mastered the material in Chapter Ten. Answer each question to the best of your ability. Correct answers and review instructions are given at the end of the test.

1. Why do earthbound observers always see the same side of the moon?

 ____

2. The moon is about what fraction of Earth in (a) diameter? __________

 (b) mass? __________ (c) average density? __________ (d) surface gravity? __________

3. Match the lunar features with their names:

 ___ (a) dry lava beds
 ___ (b) holes in the surface
 ___ (c) light-colored, higher, rugged older regions
 ___ (d) submerged clumps of mass

 1. craters
 2. highlands
 3. maria
 4. mascons

4. Suppose you were leading an expedition to explore the surface of the moon. Which of the following would be useful? (a) extra oxygen tanks; (b) flare pistol and flares; (c) flashlight; (d) magnetic compass; (e) matches; (f) star chart; (g) umbrella; (h) watch.

 ____

 Explain. ______________________________

 ____

5. What is the probable origin of most craters on the moon? __________

 ____

6. Why do the moon's surface features change so much more slowly than Earth's? ______________________________

7. How old are the (a) oldest and (b) youngest moon rocks that have been collected? ______________________________

8. Identify the three main layers of the moon.

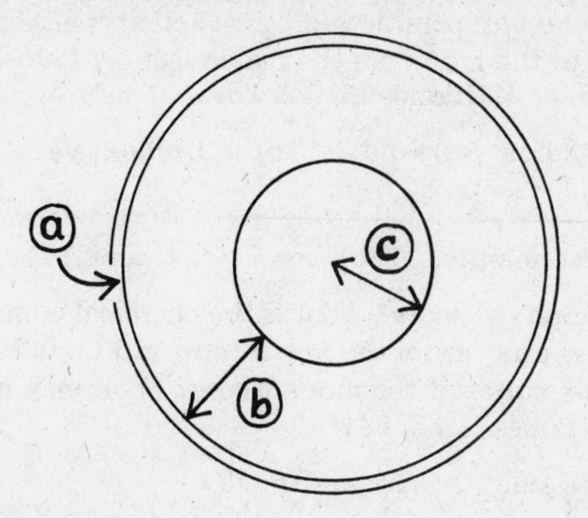

Figure 10.15. The moon's interior.

(a) ______; (b) ______; (c) ______

9. List three questions about the moon that remain to be answered.

______

______

10. What must the phase of the moon be for (a) the occurrence of an eclipse of the sun? ______ (b) an eclipse of the moon? ______

ANSWERS

Compare your answers to the questions on the self-test with the answers given below. If all of your answers are correct, you are ready to go on to the next chapter. If you missed any questions, review the frames indicated in parentheses following the answer. If you miss several questions, you should probably reread the entire chapter carefully.

1. The moon's rotation period and its period of revolution around Earth are equal, called synchronous rotation. (frame 3)
2. (a) $\frac{1}{4}$ diameter; (b) $\frac{1}{81}$ mass; (c) $\frac{3}{5}$ density; (d) $\frac{1}{6}$ surface gravity (frames 5, 6, 7)
3. (a) 3; (b) 1; (c) 2; (d) 4 (frames 8, 9, 10, 14)
4. a, c, f, h. Because the moon has no air, water, or magnetic field, nothing that requires any of these would be useful. (frames 8, 10, 14)
5. meteorites crashing into the surface (frame 9)

6. There is no air or water to cause erosion on the moon as they do on Earth. There is no comparable geological activity, either. Micrometeorites crashing into the moon are the major agents of erosion on the moon's surface. (frames 11, 12, 13)

7. (a) about 4.6 billion years old; (b) 3.1 billion years old (frame 13)

8. (a) crust; (b) mantle; (c) core (frame 14)

9. How did the moon originate? What is the chemical composition of rocks located at places away from the few Apollo and Luna landing sites? What is the true nature of the moon's core? (You may have thought of others.) (frames 6, 13, 14)

10. (a) new; (b) full (frames 16, 17)

CHAPTER ELEVEN

Comets, Meteors, and Meteorites

When beggars die, there are no comets seen:
The heavens themselves blaze forth the death
of princes.

William Shakespeare
Julius Caesar, II, ii, 30

Figure 11.1. Comet Bennett on April 16, 1970.

1. Bright *comets* have always fascinated people. Unlike ordinary stars, these fiery-looking objects appear and disappear mysteriously. Throughout history (records of bright comets go back to the fourth century B.C.) people have dreaded comets as omens of human disasters such as war or famine.

Today we know that comets belong to our solar system. They travel in elliptical orbits around the sun and follow the basic laws of physics. They are not supernatural signs at all.

(a) What was the common historical view of comets? ________________

____

(b) What is the modern astonomical view of comets? ________________

____

- - - - - - - - - - - - - - - - - -

(a) Comets were considered supernatural signs foretelling human misfortune.

(b) Comets are members of the solar system that follow natural physical laws and have no hidden meaning.

2. Comets were named for their appearance. Both the Greek word *kometes* and the Latin word *cometa* mean "long-haired." When they shine in the sky, bright comets have a starlike head, called the *nucleus*, surrounded by a glowing halo called the *coma*, with a long transparent *tail*. The nucleus is about 1 to 10 kilometers in diameter. The coma may extend 100,000 kilometers or more outside the nucleus. The tail can stream millions of kilometers into space.

Observations of comets from orbiting spacecraft above Earth's atmosphere have revealed that a hydrogen cloud, as big or bigger than the sun, surrounds a comet. This cloud is not visible from Earth.

Identify the main parts of a typical bright comet (as indicated in figure 11.2 on the following page).

(a) __________; (b) __________; (c) __________; (d) __________

- - - - - - - - - - - - - - - - - -

(a) nucleus; (b) coma; (c) tail; (d) hydrogen cloud

3. Billions of comets probably orbit far out in the solar system, but you can't see them from Earth. They shine in the sky only when they travel near the sun. The most widely accepted description of a typical comet is the *dirty snowball* model, proposed by Fred Whipple in 1950.

When a comet is far out in the solar system it consists only of a nucleus. The nucleus is probably made of frozen gases such as carbon dioxide, ammonia, methane, and water vapor (the "snow") mixed with stony or metallic particles (the "dirt").

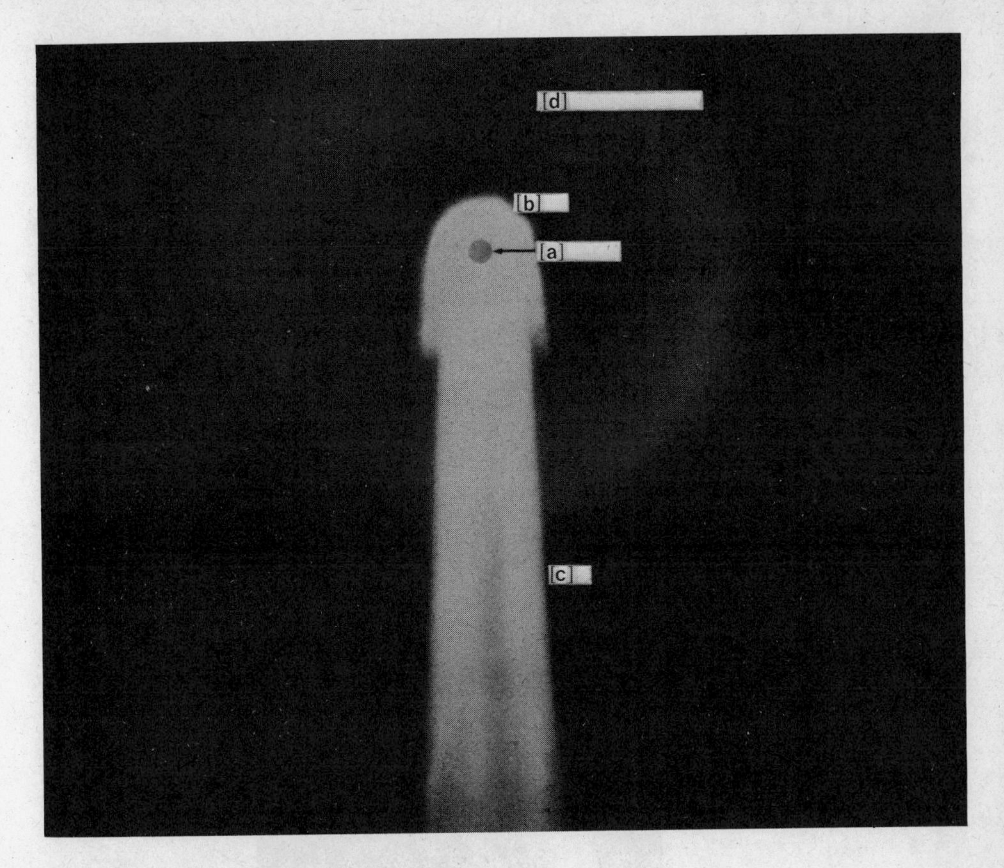

Figure 11.2. Main parts of a comet.

What is the nucleus of a comet made of? ______________________

____

— —

The dirty snowball model describes a comet nucleus as frozen gases mixed with stony or metallic particles.

4. As a comet nucleus comes in from the edge of the solar system to within a few hundred million kilometers of the sun, it warms up. Gas and dust particles evaporate from its surface. The gas and dust expand outward around the nucleus for thousands of kilometers, forming the coma. The comet shines because the dust reflects sunlight and the gases fluoresce. Astronomers using large telescopes photograph about twenty of these blurry blobs of light every year.

What causes the coma to develop? ______________________

____

— —

the sun's heat (which causes evaporation and expansion of gas and dust particles)

5. Occasionally a comet moves so close to the sun that two long, graceful comet tails form. Intense sunlight hits the dust particles (radiation pressure) and pushes them outward. The comet keeps moving, and a dust tail curves behind it. The solar wind (frame 7.14) blows ionized gases millions of kilometers straight out into space, forming a *gas,* or *ion,* tail. Comet tails are so thin that you can see right through them to stars on the other side.

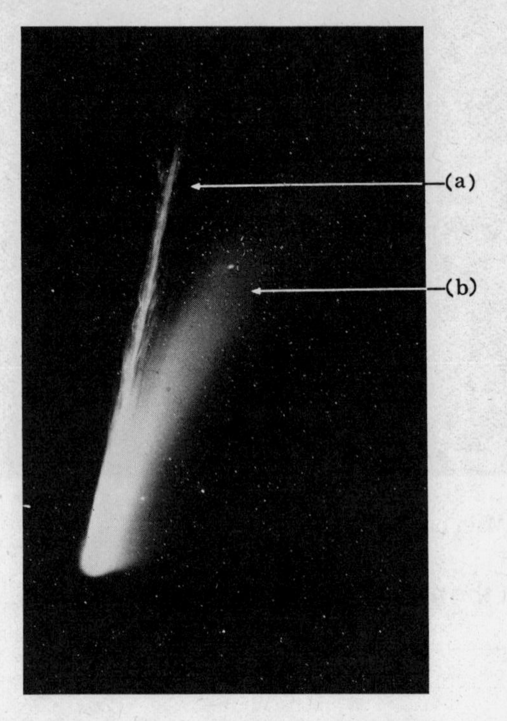

Figure 11.3. Comet with two tails.

Referring to figure 11.3, identify the dust tail and the gas (ion) tail and state the cause of each.

(a) ______________________________

(b) ______________________________

- - - - - - - - - - - - - - - - - - - -

(a) ion tail; solar wind; (b) dust tail; radiation pressure

6. If the comet goes around the sun without being completely shattered or disintegrated, it continues along its orbit back to frigid outer space. The cometary material refreezes. The coma and tail vanish.

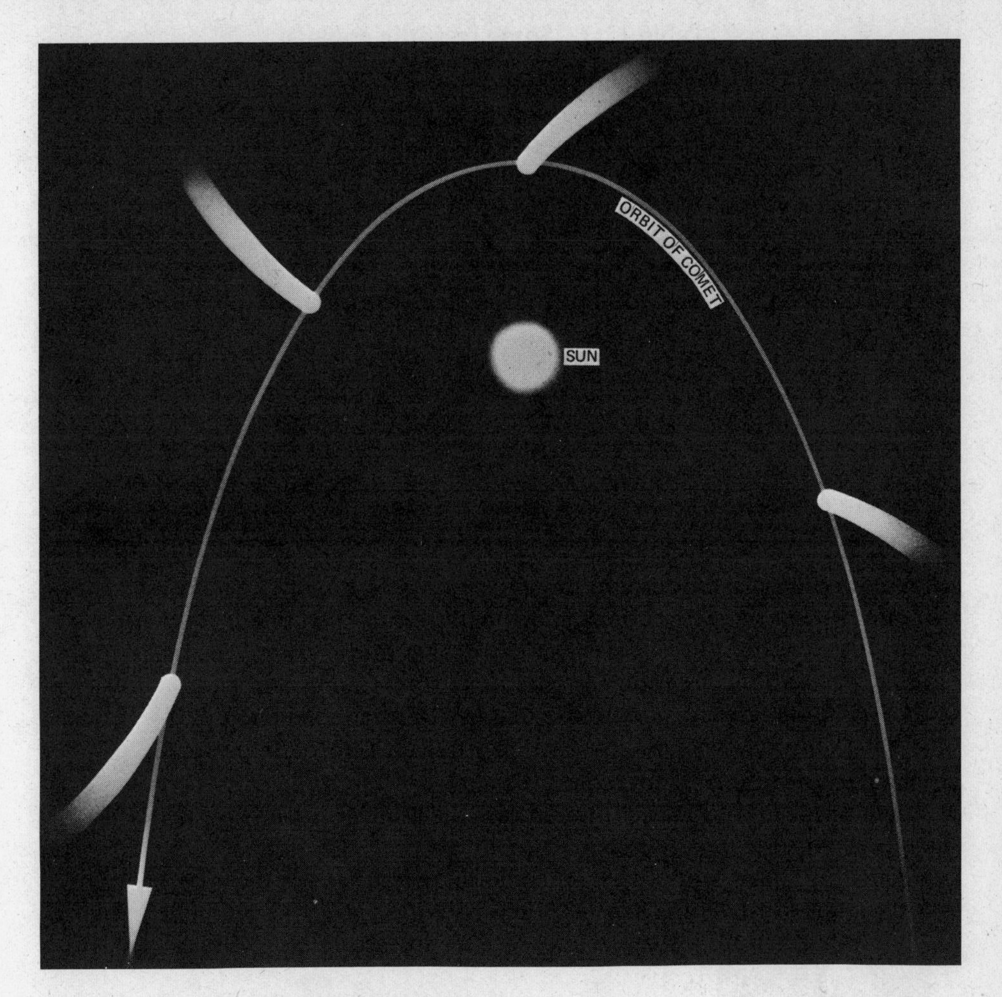

Figure 11.4. Comet's orbit around the sun.

Why do comets go back to outer space tail first? ____________________

____

- - - - - - - - - - - - - - - - - - -

Comet tails are caused by solar radiation pressure and solar wind, which are always directed away from the sun, so the tail must point away from the sun, too.

7. Astronomers have catalogued about one hundred comets that have relatively short periods of revolution around the sun. They shine periodically in the sky every time they come close to the sun. The most famous is Halley's Comet, last seen in 1910. Scientists are already making extensive plans to observe it in detail when it returns in 1986.

Table 11.1. Some periodic comets

Comet	Period (years)	Closest Approach to Sun (in AU)
Encke	3.30	0.34
Temple	5.26	1.37
Finlay	6.88	1.08
Wolf	8.42	2.50
Halley	76.1	0.59

Table 11.1 lists some comets that have appeared several times in our sky. What is the shortest known period of revolution of a comet? ____________

3.3 years (Comet Encke)

8. Figure 11.5 shows Halley's Comet on 9 different days during its last appearance in 1910.

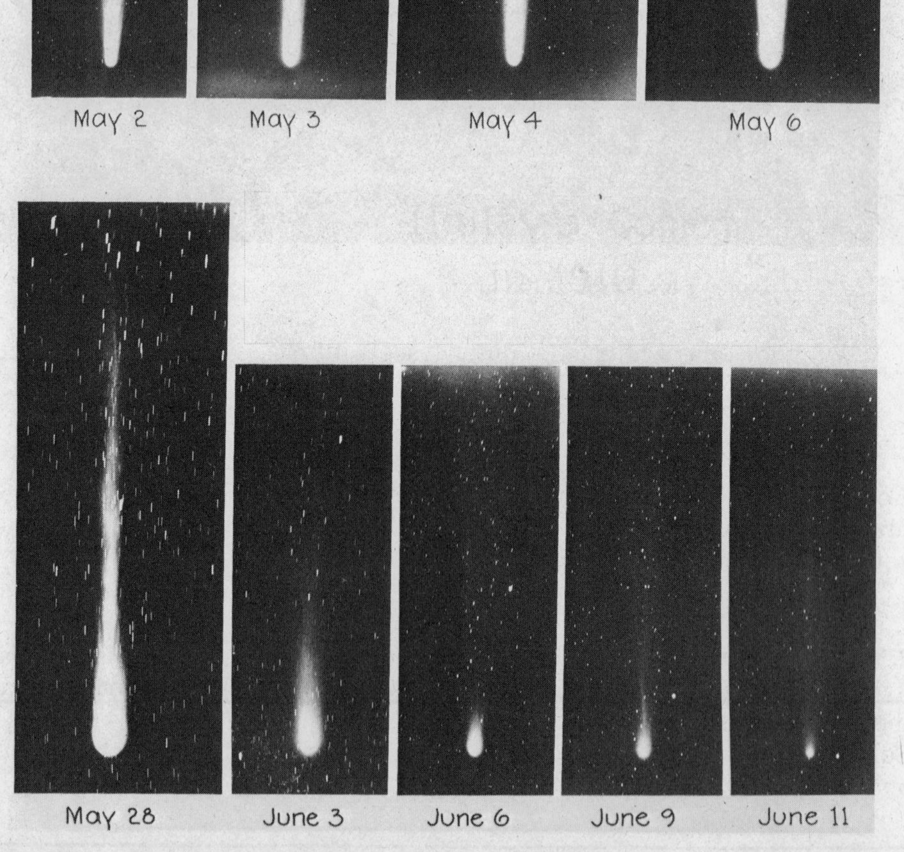

Figure 11.5. Halley's Comet on 9 different days in 1910.

Summarize five changes of appearance that a comet undergoes as it travels in its orbit around the sun. ______________________________

____

____

- - - - - - - - - - - - - - - - - - -

1. Far from the sun, the comet consists of a nucleus of frozen gases and dust. 2. Coma forms as it approaches the sun. 3. Close to the sun, the tail forms. 4. After going around the sun, cometary material refreezes. 5. Far from the sun again, coma and tail are gone.

9. Comets originate in a vast cloud of comets located 50,000 to 100,000 times farther from the sun than Earth is, according to a widely accepted theory advanced in 1950 by the Dutch astronomer Jan Oort. That Oort comet cloud, halfway out to neighbor stars, may hold 100 billion incipient comets.

Occasionally a passing star tugs on one comet, slows it down in its motion, and sends it plummeting toward the sun. That comet will be a long-period comet travelling in an ellipse around the sun unless it passes near Jupiter. Jupiter's strong gravity pulls long-period comets into new orbits.

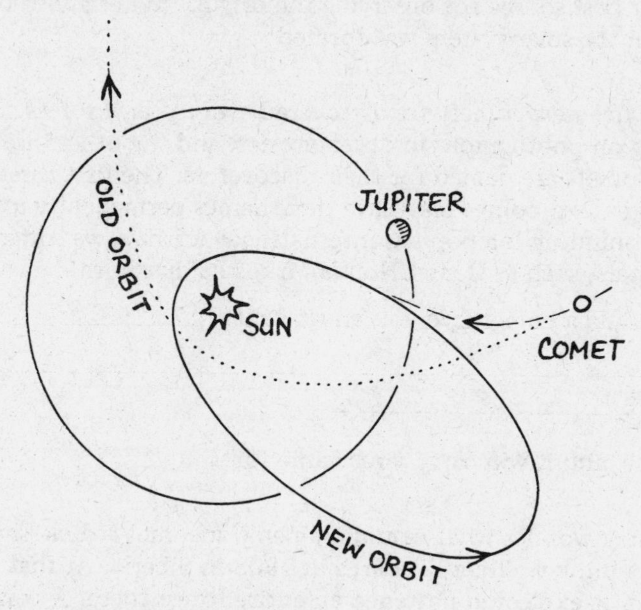

Figure 11.6. Comet moving into sun's orbit.

Where do comets probably originate? ______________________________

- - - - - - - - - - - - - - - - - - -

in a vast cloud of comets near the edge of the solar system

10. Comets that appear for the first time in our sky are important, even if they don't shine brilliantly. They are probably the only objects left that are made out of the original material from which the whole solar system formed about $4\frac{1}{2}$ billion years ago. The earth, moons, and other celestial bodies have all been changed by water, wind, or collisions. Only comets remain as they were in the beginning.

Comet Kohoutek—discovered in March 1973, nine months before its closest approach to the sun—is the best studied comet so far. Scientific observations of Comet Kohoutek were made with the most sophisticated ground-based telescopes and above Earth's atmosphere aboard Skylab (frame 7.7). They detected common metals, silicates, and compounds of carbon, nitrogen, hydrogen, and oxygen including ionized water. Ultraviolet observations showed a hydrogen cloud over 6 million miles in diameter (10 million km). (This comet is now heading toward the edge of the solar system and is not expected to visit our sky again for about 80,000 years.)

Why are new comets important? ______________________________

____

- -

They are our best source for observing the original material out of which everything in the solar system was formed.

11. About five new comets are discovered every year. Professionals find some by searching on photographs in observatories, and the others are discovered by amateurs. Comets are named for their discoverers. The first three people to report seeing a new comet may have their names permanently attached to it. Since comet-hunting is a popular international activity, we sometimes get real tongue twisters, such as Comet Honda-Mrkos-Pajdusakova!

How could a comet give you immortality? ______________________

____

- -

Discover one, and it will carry your name forever.

12. You may wonder what would happen if a comet struck Earth. A few astronomers think a comet hit Earth in 1908 in Siberia. At that time a mysterious gigantic explosion flattened an entire forest there. A comet nucleus could impact with the energy of millions of hydrogen bombs. But most astronomers figure the chances of a direct hit on Earth are extremely remote. Most comets never come anywhere near Earth in their travels around the sun.

Is it likely that Earth will be hit by a comet nucleus? ____________

- -

no

13. Each time a comet passes near the sun, some of its dust and gas are left behind along its orbit. The solid particles are called *meteoroids.* They continue to orbit around the sun like tiny planets. After many passages around the sun, all periodic comets will disintegrate completely into gas and dust.

What is a meteoroid? ______________________________

- - - - - - - - - - - - - - - - - -

a very tiny solid particle orbiting the sun out in space

14. Did you ever make a wish on a "shooting star"? These light flashes are not stars at all. They are *meteors.* Meteors are streaks of light from burning meteoroids that are plunging through Earth's atmosphere at speeds up to 72 kilometers per second. Air friction burns these tiny particles when they are 80 to 100 kilometers above Earth. On any clear dark night you can see an average of six meteors an hour flashing unpredictably in the sky. Meteors occur but are not visible during the daytime because the sky is too bright. An exceptionally brilliant meteor is called a *fireball.*

What is a "shooting star"? ______________________________

____

- - - - - - - - - - - - - - - - - -

A "shooting star," or meteor, is the streak of light that can be seen when a meteoroid is burned upon entry into Earth's atmosphere.

15. On several predictable occasions every year you may see meteors pour down from one part of the sky. This type of display is called a *meteor shower.* Meteor showers are associated with comets. They occur when Earth, moving along its orbit around the sun, crosses a swarm of meteoroids left behind by a comet.

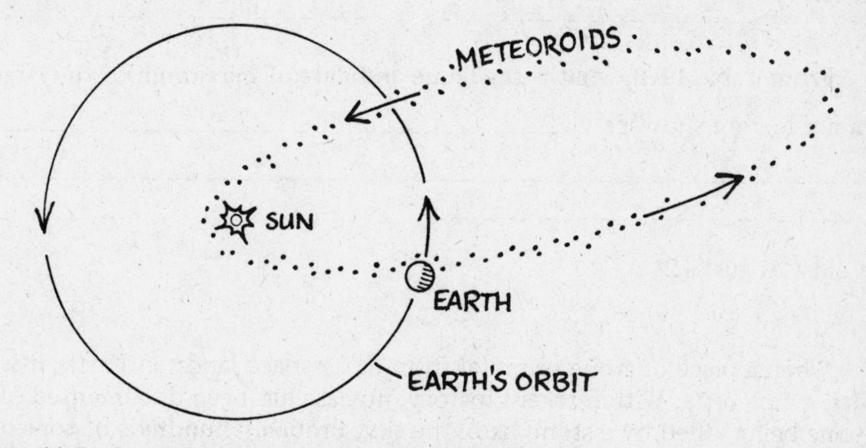

Figure 11.7. Conditions that create a meteor shower.

In 1910, when Earth was about to travel right through the tail of Halley's Comet, people panicked. What would you expect to happen if Earth passed through a comet's tail? ______________________________

- -

a meteor shower

16. During a meteor shower, all the meteors look like they come from one common point in the sky called the *radiant* of the shower. Meteor showers are named for the constellation where they seem to originate. You can usually see more meteors after midnight than before, because Earth, moving along its orbit, is then travelling head-on into the swarms of particles. Meteor showers are best seen with your naked eye on nights when the moon is not bright. A full moon can blot out a meteor shower.

Table 11.2. Principal annual meteor showers

Shower	Date of Maximum	Approximate Hourly Rate	Associated Comet
Quadrantids	January 3	30	
Lyrids	April 23	8	1861I
Eta Aquarids	May 4	10	possibly Halley
Delta Aquarids	July 30	15	
Perseids	August 12	40	1962III
Orionids	October 21	15	possibly Halley
Taurids	November 4	8	Encke
Leonids	November 16	6	1866I Temp
Geminids	December 13	50	
Ursids	December 22	12	Tuttle

From Table 11.2, what is the name and date of maximum of the largest summer meteor shower? ______________________________

____

- -

Perseids; August 12

17. When a piece of stone or metal from outer space lands on Earth, it is called a *meteorite*. Within recent history, no case has been documented of anyone being killed by a stone from the sky. Probably hundreds of tons of cosmic material fall through the atmosphere every year, but only about two or three meteorites per decade land in places where people live, and even more rarely is any injury reported.

On March 8, 1976, a red fireball the size of a full moon was seen by scores of people in the northeast China province of Heilungkiang. A violent break-up occurred when the fireball was 17 kilometers over the northeastern part of Kirin City. Tens of thousands of people over a wide area observed the entry of the meteor into the earth's atmosphere and its descent and explosive impact with the ground. Large and small stones and fragments have been turned over to scientists for study.

What is a meteorite? ____

---

a piece of stone or metal from outer space

18. If you need extra spending money, find a meteorite! Scientists have paid $100 for genuine material from outer space. Right now meteorites are the only extraterrestrial material (from outer space), besides the moon rocks and soil from the Apollo and Luna trips, that scientists can study close up.

Meteorites are divided into three main types. (1) *Iron* meteorites are about eight times as dense as water, and are mostly iron (about 90 percent) and nickel. (2) *Stony-irons* are about six times as dense as water. Their estimated age is 4.6 billion years, which is the approximate age of the whole solar system. They contain iron and nickel and silicates. (3) *Stones* are about three times as dense as water. They have a high silicate content, and only about 10 percent of their mass is iron and nickel.

Stony meteorites look like ordinary Earth rocks. They are not usually noticed unless they are seen falling. Table 11.3 lists the occurrence of the different types of stone and iron meteorites. The "seen falling" column probably represents true relative abundance.

Table 11.3. The occurrence of meteorite types

Types	Meteorites Seen Falling	Meteorite Finds
Irons	6%	66%
Stony-irons	2%	8%
Stones	92%	26%

In 1969 a rare kind of stony meteorite with a high carbon content—called a *carbonaceous chondrite*—fell near Murchison, Victoria, Australia. It contained simple amino acids, the building blocks of life.

Recently a large collection of uncontaminated meteorites was recovered from the ice in Antarctica. Some of the 4.5 billion year old meteorite specimens are carbonaceous chondrites, containing amino acids. Since these finds show that prebiotic matter formed beyond Earth, they add excitement to the search for life in space (more on this in Chapter 12).

The largest meteorite ever found—the Hoba West—weighs about 66 tons. It still lies in South-West Africa where it landed. Meteorites are usually named after the post office closest to their landing site.

Figure 11.8. The Barringer crater.

Large meteorites carve out huge craters in the earth when they crash. You can see the most famous meteorite crater in the United States, the Barringer, near Winslow, Arizona. It is about 1.5 kilometers across and 180 meters deep. Some of the largest meteorites on display in the United States are listed below.

Meteorite	Approximate Weight	Present Location
Ahnighito (Greenland)	34 tons	American Museum of Natural History (New York City)
Willamette (Oregon)	14 tons	American Museum of Natural History
Furnas County (Nebraska)	1 ton	University of New Mexico
Paragould (Arkansas)	800 pounds	Chicago Natural History Museum

Why are meteorites important? ________________________________

____

- - - - - - - - - - - - - - - - -

They are extraterrestrial material that scientists can examine closely to determine more about the solar system.

SELF-TEST

This self-test is designed to show you whether or not you have mastered the material in Chapter Eleven. Answer each question to the best of your ability. Correct answers and review instructions are given at the end of the test.

1. What is a comet nucleus made of? ______________________________

____

2. Describe five changes of appearance that a comet undergoes as it travels in its orbit around the sun. ______________________________

____

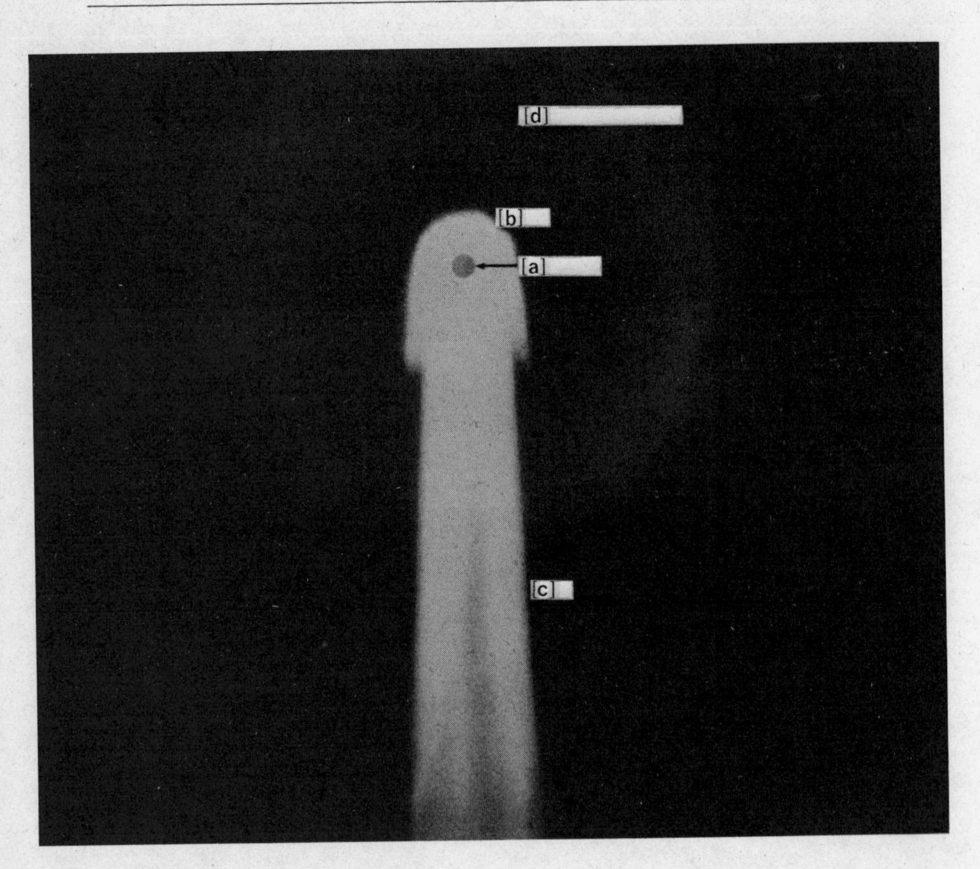

Figure 11.9. Main parts of a comet.

3. Identify the main parts of a comet. (a) ______________; (b) ______________; (c) ______________; (d) ______________

4. Match each description with the correct item.

    ___ (a) "shooting star"
    ___ (b) small particle orbiting the sun
    ___ (c) solid body that reaches Earth

    1. meteor
    2. meteorite
    3. meteoroid

5. Explain the relationship between comets and meteor showers. ___

6. Describe the composition and probable origin of meteorites. ___

7. List the following in order of increasing distance from the sun: asteroid belt, Earth, Oort comet cloud, Pluto.

8. Explain why comets and meteorites are of interest to scientists.

ANSWERS

Compare your answers to the questions on the self-test with the answers given below. If all of your answers are correct, you are ready to go on to the next chapter. If you missed any questions, review the frames indicated in parentheses following the answer. If you miss several questions, you should probably reread the entire chapter carefully.

1. A comet nucleus probably consists of frozen gases such as methane, ammonia, carbon dioxide, and water vapor mixed with stony or metallic particles, according to the dirty snowball model. (frame 3)
2. Far from the sun, a comet consists of a nucleus of frozen gases and dust. Coma forms as it approaches the sun. Tail forms close to the sun. After going around the sun, comet refreezes. Far away from the sun, comet consists of a nucleus again. (frames 3, 4, 5, 6, 8)
3. (a) nucleus; (b) coma; (c) tail; (d) hydrogen cloud
    (frame 2)
4. (a) 1; (b) 3; (c) 2
    (frames 13, 14, 17)

5. Meteor showers occur when Earth, moving along its orbit around the sun, crosses a swarm of meteoroids left behind in space by a comet. (frame 15)

6. iron meteorites—mostly iron (about 90 percent) and nickel; stony irons—iron and nickel and silicates; stones—high silicate content, only about 10 percent iron and nickel by mass; probable origin: asteroid belt (frame 18)

7. Earth, asteroid belt, Pluto, Oort comet cloud (frames 9, 18)

8. Because they originate in space and can help us understand the history and composition of our own planet, Earth, and of the solar system. (frames 10, 17, 18)

CHAPTER TWELVE

Life on Other Worlds?

> **We hope someday, having solved the problems we face, to join a community of galactic civilizations. This record represents our hope and our determination, our good will in the vast and awesome universe.**
>
> **President Jimmy Carter, 1977, recording aboard Voyager**

1. No one knows whether *extraterrestrial life*—life off Earth—exists. But the search has begun!

Two American Viking (robot) spacecraft began the first quest for life on another planet, Mars, in 1976. Radio astronomers in the United States and the Soviet Union are probing the sky for intelligent signals from other civilizations. The American effort is called SETI, or *s*earch for *e*xtra*t*errestrial *i*ntelligence.

Biochemists say all living organisms on Earth depend on a few basic molecules which they can manufacture from gas atoms in the laboratory. Astronomers find these basic atoms and molecules of life on other planets, in stars, and in interstellar dust clouds (see frame 5.4). They have also found amino acids on stony meteorites (see frame 11.18). Physicists assume that the natural laws that rule physical and chemical events on Earth apply everywhere in the universe.

If life on Earth evolved from nonliving molecules in a series of physical and chemical reactions and is not a unique cosmic accident, then life probably developed elsewhere in the universe, too.

Explain why many scientists think extraterrestrial life may exist.

____

____

– – – – – – – – – – – – – – – – – –

The basic molecules of life have been found in space and manufactured in the laboratory. If living organisms evolve from nonliving molecules in a series of physical and chemical reactions and are not a unique cosmic accident, then life may have occurred on other worlds.

2. The most basic characteristics that separate a *living organism* from a nonliving one are the ability to reproduce and metabolism. What causes the spark of life? No one knows, but the *cosmic evolution* theory ties the appearance of living organisms to universal forces in the following way:

About 4 billion years ago, Earth's atmosphere was mainly compounds of hydrogen, carbon, oxygen, and nitrogen gases. A mixture of the noxious gases, methane, ammonia, and carbon dioxide, and water blanketed our planet. The sun's cosmic rays and lightning flashes provided energy to bind these gases into more complex *organic molecules* (molecules containing carbon) such as amino acids, the basic molecules of life.

Gradually, organic molecules accumulated in Earth's seas. As their concentration increased, collisions between molecules in the water joined small molecules into larger ones. Water was important in this process, because it speeds chemical reactions by facilitating collisions between molecules.

A billion years passed as more and more complex molecules formed. Eventually molecules were formed that were capable of reproduction and metabolism, and the threshold from nonliving to living matter was crossed.

Why is water important in the chemical evolution of the basic organic molecules of life? ____

____

- - - - - - - - - - - - - - - - - - - -

Water speeds up chemical reactions by enabling molecules to collide with each other.

3. The common *virus* seems to confirm that living organisms can evolve out of nonliving molecules, because it has characteristics of both.

Viruses are submicroscopic objects that can cause diseases in animals or plants. A virus is the simplest object that can be called living, because it can reproduce when in contact with living cells.

Viruses range in size from smaller than many nonliving molecules to as large as a small bacterium. They are sometimes regarded as complex nonliving proteins since they lack an energy source and the raw materials needed for growth.

What evidence exists that living organisms may evolve out of nonliving molecules? ____

- - - - - - - - - - - - - - - - - - - -

The virus has characteristics of both living organisms and nonliving molecules.

4. The principle of *natural selection*, or the *survival of the fittest*, is a theory that asserts that all living creatures on Earth evolved from simple one-celled organisms.

Fossil records in ancient rocks show that life existed on Earth as simple one-celled plants called algae and organisms such as bacteria over 3 billion years ago.

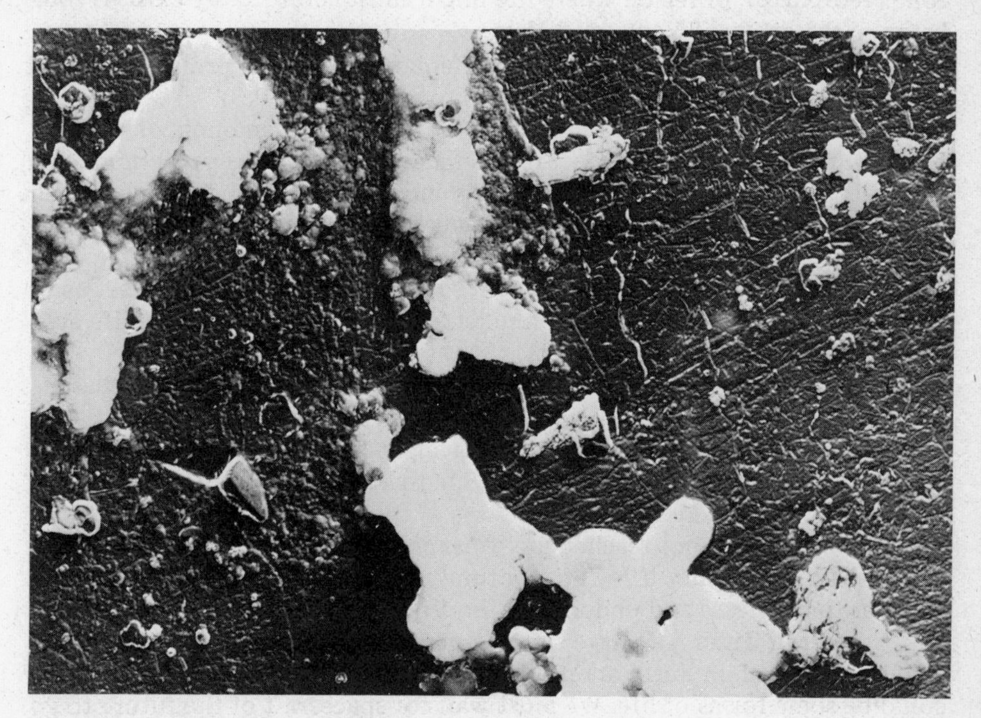

Figure 12.1. Fossil bacteria, several billion years old.

The first living species reproduced. But children are never exact copies of their parents. Some *variations* in characteristics are always introduced whenever reproduction takes place.

Those individuals with favorable variations that helped them to survive the environment were the fittest. They had the best chance of reaching adulthood and reproducing themselves. Thus favorable traits were passed on, while unfavorable traits died out, by natural selection. Slowly over a long period of time a new species could have formed from the original one.

In this way, over billions of years under the varying environmental conditions that existed on Earth, all living things, including human beings, may have evolved from simple cells.

Suggest an environmental pressure that could produce a critical evolutionary advance. ______________________________

- - - - - - - - - - - - - - - - - - - -

a drastic world-wide change in climate. (You may have thought of others.)

5. Life may also have evolved on a neighboring planet. The sun's *habitable zone* (ecosphere), or region where life might most comfortably exist, is roughly between the orbits of Venus and Mars.

Venus does not seem suitable for life, because it has a prohibitively hot surface temperature up to 900° F and is too dry (frame 9.3).

Mars seems more suitable. It looks as though large amounts of water, essential for Earth-life, once occurred on Mars. Photographs show branching channels that look like familiar riverbeds and tributaries, where mighty rivers may once have flowed. Spacecraft data show water in the Martian polar ice caps, frost, fog, and filmy clouds (frame 9.11 and figure 9.16).

Experiments on Earth have shown that some plants and microbes can survive environmental conditions similar to those on Mars today. If life ever evolved there, perhaps it still exists.

The Viking Lander experiments were designed to detect carbon-based microbes living in the Martian soil. The results are puzzling. There was definitely some form of activity there. It might have been due to living organisms, but it might also have been due to some unusual chemical characteristic of the Martian soil.

Biologists have not reached any final conclusions about the presence or absence of life on Mars. They are continuing experiments on Earth in an attempt to duplicate and understand the Viking results.

Jupiter or Titan may have simple microbes. The gases out of which life probably evolved on Earth are present there. Beneath the cloud cover, there could be some forms of life. We must wait for spacecraft of the future to go and find out.

Give three observations that suggest life may have developed from nonliving molecules on Mars. ______________________________

____

- - - - - - - - - - - - - - - - - - - -

(1) Evidence indicates that a lot of water once flowed on the planet.
(2) Mars is inside the sun's habitable zone.
(3) Some forms of earth-life can survive Martian conditions.

6. We may be the only *intelligent* (as we proudly call ourselves) *civilization* in the entire universe. Or many others may exist.

Our sun is just one of about 100 billion stars in our Milky Way Galaxy.

Roughly a billion galaxies are within sight of our largest telescopes. Many of the stars in these galaxies may have life-supporting planets circling them. And some of these planets may bear intelligent civilizations.

Scientists such as Carl Sagan, Frank Drake, and I.S. Sklovsky *estimate* the *number of intelligent civilizations* in our own Milky Way Galaxy—the only ones we could hope to communicate with at present—by this method: They estimate with some confidence (1) the number of stars in the Galaxy, (2) the fraction of those stars that have planets circling them, and (3) the average number of planets suitable for life.

With much less confidence, they estimate (4) the fraction of suitable planets where life has actually developed, (5) the fraction of those life starts that evolve to intelligent organisms, and (6) the fraction of intelligent species that have attempted communication.

Then they just guess (7) the average lifetime of an intelligent civilization.

When all these factors are considered, estimates range anywhere from one intelligent civilization (ours) up to a million of them around the Milky Way Galaxy today.

Why do you think (7), the average lifetime of an intelligent civilization, is the most uncertain number of all? ______________________________

____

- - - - - - - - - - - - - - - - - - -

No one knows what happens when a civilization like ours, if we are typical, reaches a stage of technical development where it can communicate with other civilizations in our Galaxy. Will it last long enough for a conversation, or will it self-destruct with nuclear weapons, pollution, or overpopulation?

7. The condensation theory of the formation of stars asserts that many stars must be orbited by planets (frame 7.3). Astronomers at the Sproul Observatory in Swarthmore, Pennsylvania, may have discovered planets circling our neighbor, Barnard's Star.

Planets belonging to other stars cannot be seen even with the largest telescopes available. Planets, if they exist, are lost in the glare of their star. Astronomers search for invisible companions by observing the proper motion (frame 3.9) of the visible star.

Peter Van de Kamp has reported a wobble in the proper motion of Barnard's Star which has been photographed from 1937 on. The wobble could be caused by the gravitational tug of planets. The data suggest that there are two massive planets orbiting Barnard's Star. Astronomers are uncertain, because the observations are very difficult to make.

What does the study of proper motion of Barnard's Star suggest to scientists? _______________________________________

- - - - - - - - - - - - - - - - - - -

The star's wobble may indicate the presence of planets.

8. *Interstellar travel*, trips to other stars, would be the most dramatic way to see if other civilizations exist. But no one is seriously planning a deep space voyage today for several reasons.

Even the closest stars are several light-years away, and no spacecraft can travel anywhere near the speed of light. A trip to our nearest neighbor, Alpha Centauri, at the speed the Apollo astronauts travelled to the moon, would require thousands of years.

If we could build spacecraft to travel near the speed of light, more energy than the whole world now uses in a hundred years would be required for a round trip to a nearby star.

A long space voyage and prolonged weightlessness could injure the crew. The Skylab 4 astronauts lived in space for the American record of 84 days, 1 hour, 16 minutes in 1974. *Space shuttle* crews may orbit Earth for up to a month at a time. Soviet cosmonauts lived aboard the orbiting *Salyut 6* space station for a world record 185 days in 1980. Their physiology and body chemistry were changed by weightlessness. Research continues on the biomedical effects of space flight and ways to help people adapt to long-term weightlessness.

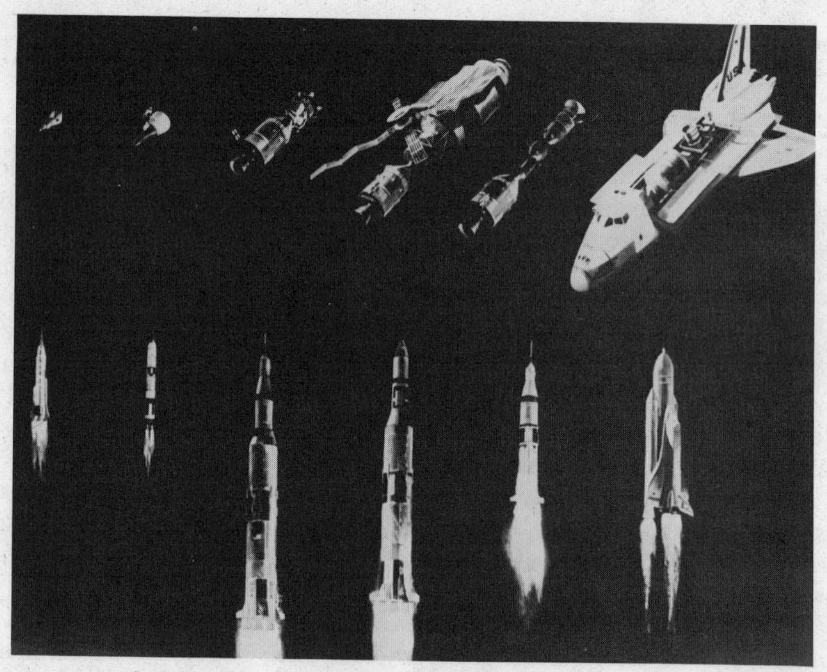

Figure 12.2. NASA manned missions. Left to right: Mercury/Atlas; Gemini/Titan II; Saturn V/Apollo; Saturn V/Skylab; Saturn IB/Apollo, Space Shuttle/Orbiter.

The robot spacecraft Pioneer 10 which flew past Jupiter in 1973–1974 will be the first ever to leave our solar system, in 1987. Pioneer 10 and 11 each bear a plaque carrying a symbolic message from us to any intelligent beings they encounter. The plaque is intended to show when, where, and by whom each spacecraft was launched.

The Voyager spacecraft launched in 1977 (frame 8.12) carry a unique

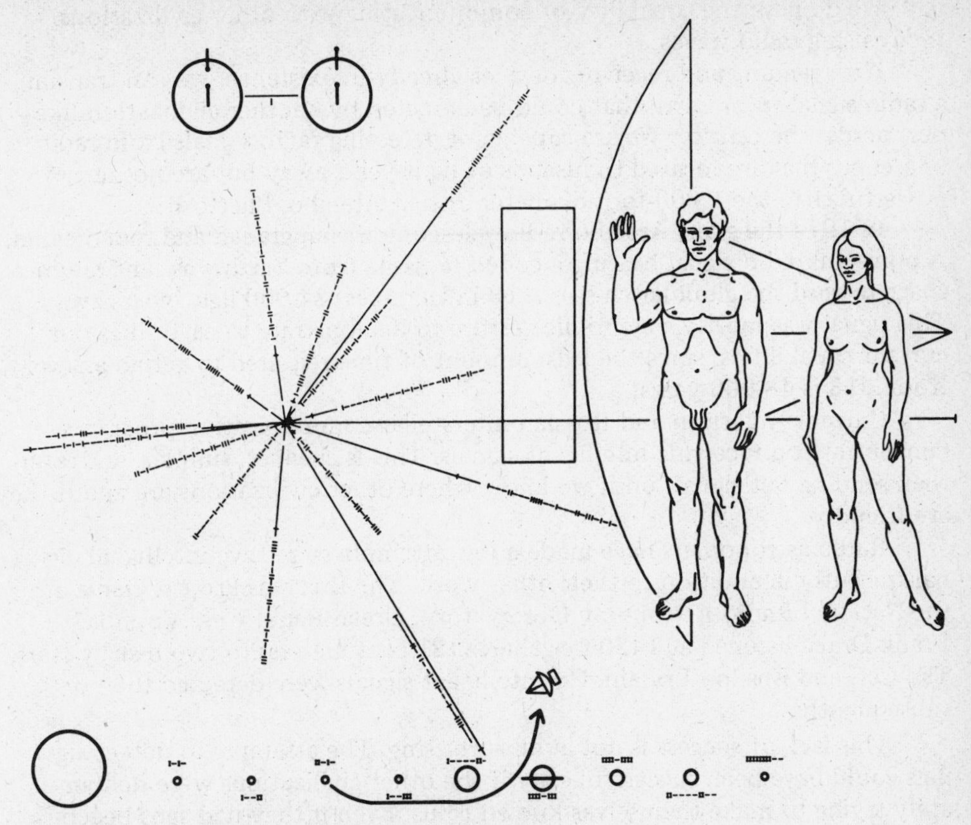

Figure 12.3. The plaque aboard Pioneer 10 and 11.

record (plus cartridge) with information encoded electronically about earthlings. Possible civilizations out in space can hear recordings of birds, music, and greetings in sixty languages. President Jimmy Carter recorded a good-will message whose excerpt is included as the opening quotation of this chapter.

At present, to send large numbers of high-speed unmanned spacecraft to probe other stars would be enormously expensive. Perhaps in the future unmanned probes will be sent to a few nearby stars for scientific exploration.

Most scientists do not think we have been visited by creatures from other worlds, either, although you often hear accounts of UFO sightings. They would be glad to examine concrete evidence, such as a piece of the spacecraft, in the laboratory, but none has ever been found.

Why is it unlikely that Earth will be invaded by hostile beings from space, as some alarmists have suggested? ______________________________

____

— — — — — — — — — — — — — — — — —

Travel across the enormous distances between stars would take much too long and exhaust too many resources to be justifiable for most purposes, if other civilizations are at our stage of development.

9. We do have the capability of communicating with other civilizations today using radio waves.

With sending and receiving devices already in existence, we can transmit a radio signal from Earth that could be detected by another civilization like ours across the Galaxy. We are capable of detecting radio signals from radio telescopes that are located thousands of light-years away but are no more powerful than the 1,000-foot-diameter dish at Arecibo, Puerto Rico.

In 1974 the giant Arecibo radio telescope was upgraded and rededicated. A powerful radio signal bearing a coded message from Earth was sent out into space toward the globular cluster M13 in Hercules, 24,000 light-years away. This signal was mostly a symbolic gesture to demonstrate to earthlings our current capabilities, since the least amount of time required to get an answer from M13 is 48,000 years!

Current American and Russian interstellar communication programs concentrate on receiving intelligent signals. This is cheaper, simpler, and safer than sending out signals until we know where other civilizations are and if they are friendly.

Radio astronomers have made a few attempts to receive intelligent signals recently in addition to their other work. The first was Project Ozma at the National Radio Astronomy Observatory, Green Bank, West Virginia. Frank Drake listened at 1420 megahertz (21 centimeters) to two nearby stars, Tau Ceti and Epsilon Eridani. No intelligent signals were detected then or subsequently.

This lack of success is not at all surprising. The attempts to detect signals could have been successful only if the other civilizations were deliberately trying to make themselves known to us. Even if they had sent beacons to us, we might not have looked in the right direction at the right time, tuned to the right frequency. The process is rather like trying to find a needle in a haystack by looking only occasionally, and without even knowing what a needle looks like.

The exciting search for extraterrestrial life has really only just begun.

About how long would it take a radio-wave message from Earth to reach a planet circling Alpha Centauri (4.3 light-years away)? Hint: Radio waves travel at the speed of light (frame 2.5). ______________________________

- - - - - - - - - - - - - - - - - - - -

4.3 years

10. The first detailed proposal for a highly sensitive search for other intelligent civilizations—Project Cyclops—was made in 1971.

A fairly thorough search for extraterrestrial intelligent life could be conducted for a cost of a few billion dollars by using a vast array of some one thousand radio telescopes together with complex computers.

Recently scientists reexamined the concept of interstellar communication as part of a SETI feasibility study, conducted by NASA.

They recommended that we now begin a serious international search for extraterrestrial intelligence with the United States taking a lead.

Figure 12.4 A schematic drawing of the massed radio telescopes of Project Cyclops.

Large systems involving expensive construction are not necessary for a significant SETI program today. We could, at a modest cost, look for signals from the vicinity of all known stars similar to our sun within 100 light-years of Earth using existing radio telescopes and computers.

The frequencies considered most likely for our first contact lie between 1400 and 1730 megahertz, often referred to as the galactic "waterhole" where we shall all meet.

In the future if we decide to undertake a more comprehensive search for beacons coming from 1,000 light-years away, we can construct large systems such as Project Cyclops. Alternatives such as putting antennas in space or on the moon are also possibilities for the future.

Imagine that as a citizen you have been asked to cast a vote for or against our joining a serious international search for signals from other civilizations.

How would you vote, and why? ______________________________

____

\- - - - - - - - - - - - - - - - - -

Your answer is a matter of opinion. I would vote for the search. If we found other intelligent civilizations, they might teach us how to surmount the problems now threatening our survival on Earth. If we did not, the money would still be well spent, since we could expect tremendous gains in knowledge for mankind's benefit from an intellectual and scientific effort of this size.

SELF-TEST

This self-test is designed to show you whether or not you have mastered the material in Chapter Twelve. Answer each question to the best of your ability. Correct answers and review instructions are given at the end of the test.

1. List two observations that support the theory that the first living things on Earth may have evolved spontaneously out of non-living chemicals.

   ____

   ____

2. Briefly summarize the scientific theory of the evolution of intelligent life on Earth from simple one-celled organisms. ______________

   ____

   ____

3. Explain why the first search for life on another planet was conducted on Mars. ______________________________

   ____

4. Give the evidence for planets circling Barnard's Star. ______________

   ____

5. What is the most uncertain factor involved in estimates of the statistical chances for extraterrestrial intelligent life? ______________

6. What is the dominant current scientific view of UFOs? ______________

   ____

7. Match the following firsts to the correct project. First:

   ___ (a) American effort to receive intelligent signals from outer space

   ___ (b) message to distant star cluster

   ___ (c) possible future thorough search for extraterrestrial life using massed array of radio telescopes and high-speed computers

   ___ (d) search for life on surface of another planet

   ___ (e) spacecraft to leave solar system, bearing a message from Earth

   1. Arecibo radio telescope
   2. Pioneer 10
   3. Project Cyclops
   4. SETI
   5. Viking 1 and 2

ANSWERS

Compare your answers to the questions on the self-test with the answers given below. If you missed any questions, review the frames indicated in parentheses following the answer. If you missed several questions, you should probably reread the entire chapter carefully. Look in the Appendix for other interesting and useful information about astronomy.

1. (a) Biologists say all living organisms on Earth depend on a few basic molecules. These molecules have been manufactured from gas atoms in the laboratory. (2) The common virus has characteristics of both living organisms and non-living molecules. (frames 1, 3)
2. The principle of natural selection, or the survival of the fittest, asserts that when living species reproduced, variations in characteristics were introduced. Those individuals with favorable variations that helped them to survive the environment were the fittest. They had the best chance of reaching adulthood and reproducing themselves. Thus favorable traits were passed on while unfavorable traits died out, by natural selection. Intelligence is a favorable trait. Over a period of millions of years under the different environmental conditions that existed on Earth, intelligent beings evolved from simple cells. (frame 4)
3. Mars is in the sun's habitable zone; evidence suggests that water once flowed on Mars; certain terrestrial plants and microbes can survive environmental conditions similar to those on Mars; and Mars is close enough for the trip to be cost-effective. (frames 5, 8)
4. The wobble in the proper motion of Barnard's Star may be caused by the gravitational tug of massive planets. (frame 7)
5. the average lifetime of an intelligent civilization (frame 6)
6. Most scientists do not think we have been visited by creatures from other worlds. They would be glad to examine concrete evidence, such as a piece of the spacecraft, in the laboratory, but none has ever been found. (frame 8)
7. (a) 4; (b) 1; (c) 3; (d) 5; (e) 2 (frames 1, 8, 9, 10)

Epilogue

> Astronomy compels the soul to look upwards
> and leads off from this world to another.
>
> Plato (c. 428–348 B.C.)
> *The Republic*

Astronomy has come a long way since the ancients first pondered the mysteries of the universe. But many exciting discoveries still lie ahead. Now that you have deepened your understanding of basic ideas, you should enjoy observing the sky and contemporary discoveries more than ever!

Behavioral Objectives

A precise set of behavioral or instructional objectives has been prepared to assist you in learning the material of this book. These objectives indicate what you should be able to do once you have mastered the book's contents. When you have completed this book you will be able to do the following things.

CHAPTER 1

- Locate sky objects by their right ascension and declination on the celestial sphere.
- Identify some bright stars and constellations visible each season.
- Explain why the stars appear to move along arcs in the sky during the night.
- Explain why some of the constellations which appear in the sky each season vary.
- Explain the apparent daily and annual motions of the sun.
- Define the zodiac.
- Describe how the starry sky looks when viewed from different latitudes on Earth.
- Define a sidereal day and a solar day, and explain why they differ.
- Explain how astronomers classify objects according to their apparent brightness (magnitude).
- Explain why the Pole Star and the location of the vernal equinox change over a period of thousands of years.

CHAPTER 2

- Describe the wave nature of light, including how it is produced and travels.
- Name the major regions of the electromagnetic spectrum from shortest wavelength to longest.
- State the relationship between wavelength and frequency.

- State the relationship between the color of a star and its temperature.
- List the three windows (wavelength bands) in Earth's atmosphere in order of their importance to observational astronomy.
- Explain how refracting and reflecting telescopes work.
- Define light-gathering power, resolving power, and magnification for a telescope.
- State the two factors that are most important in telescope performance.
- State the purpose of a spectrograph.
- Explain how radio telescopes work, and list some interesting radio sources.
- Explain why ultraviolet, X-ray, and gamma ray telescopes must operate above Earth's atmosphere, and list some objects they study.
- Explain why infrared telescopes are located in very high dry sites and list some objects they observe.

CHAPTER 3

- Describe the method and limitations of the parallax method to determine the distance to stars.
- Describe three types of spectra: emission, absorption, and continuous spectra.
- Explain why emission and absorption spectra are unique for each element.
- Give a general description of stellar spectra, and explain how they are divided into spectral classes.
- Explain how a star's chemical composition, surface temperature, and radial velocity are determined from its spectrum.
- List several other kinds of information that are obtained from stellar spectra.
- Explain how a star's proper motion and space velocity are determined.
- Explain the difference between apparent brightness and absolute luminosity.
- Explain the relationship between apparent magnitude, absolute magnitude, and distance.
- Describe the H-R diagram, including an explanation of the relationship of a star's mass to its luminosity and temperature.
- Compare red giants and white dwarfs with our sun, in terms of mass, diameter, and density.
- Define four classes of binary stars.

CHAPTER 4

- Define stellar evolution.
- List the stages in the life cycle of a star like our sun according to the modern theory of stellar evolution.

- Explain the importance of the H-R diagram to theories of stellar evolution.
- Explain the relation between a star's age and its position on the H-R diagram.
- List the three main steps in the birth of a star.
- Describe the energy balance and pressure balance in main sequence stars.
- Compare and contrast what happens in the advanced stages of evolution for stars of large and small mass: planetary nebulas, white dwarfs, supernovas, pulsars and neutron stars, black holes.
- Identify nebula (birthplace of stars), main sequence star, blue giant, red giant, and pulsating variable stars that can be observed in the sky.
- Describe the origins of the different chemical elements and the importance of supernovas to new generations of stars.
- Describe observational evidence for supernovas, neutron stars, and black holes.

CHAPTER 5

- Define a galaxy and a cluster of galaxies.
- State the shape, size, number of stars, and appearance of the Milky Way Galaxy and give the location of the sun and Earth within it.
- Describe the contents of the interstellar medium.
- Compare and contrast emission and dark nebulas.
- Compare and contrast open (galactic) and globular clusters.
- Explain how radio maps of our Galaxy are constructed.
- Explain the method of using H-R diagrams to determine the ages of star clusters.
- Distinguish between population I and II stars.
- Identify the most distant object visible to the naked eye.
- Compare and contrast properties of the three main types of galaxies in the Hubble classification scheme.
- Describe differences between a normal galaxy and each of the following: exploding galaxies, colliding galaxies, radio galaxies, Seyfert galaxies, and quasars.

CHAPTER 6

- Define cosmology.
- Describe the limitations of cosmology.
- Describe the evidence that the universe is expanding.
- State the Hubble law.
- Explain the meaning of the Hubble constant.
- Describe the past, present, and future of the universe according to each of the three main cosmological models—big bang, oscillating, and steady state theories.

- Describe four methods for choosing among the big bang, oscillating, and steady state theories.
- List observations in support of the evolutionary theories.

CHAPTER 7

- List some reasons why modern astronomers study the sun.
- Define the solar constant, and explain why it is important to know if it is truly constant with time.
- Define the astronomical unit (AU).
- Describe the sun as a star: formation, properties, motion.
- Describe the structure of the sun: core, zone of convection, photosphere, chromosphere, and corona.
- Describe the sun's rotation and magnetic field.
- List the basic physical dimensions of the sun.
- Describe some modern tools and techniques for studying the sun.
- Describe the origin, properties, and the cyclic nature of sunspots, and describe how sunspot variations are related to solar activity.
- Describe the location, origin, and nature of solar granules, faculae, plages, prominences, and flares.
- Describe the origin and nature of the solar wind.
- Outline the puzzle of the missing solar neutrinos.

CHAPTER 8

- List the members of the solar system.
- State the essential difference between a planet and a star.
- Describe evidence supporting the condensation theory of the formation of the solar system.
- Explain the phases of the moon.
- Describe the development of our understanding of the solar system, including the contributions of Ptolemy, Copernicus, Galileo, Tycho Brahe, Kepler, and Newton.
- State and apply the laws governing the motions of bodies under gravity.
- Explain the apparent motions of the planets, including retrograde motion.
- Explain why the moon's sidereal month and synodic month differ.
- Differentiate between the revolution and rotation of celestial bodies.
- Explain the motions of Earth-orbiting satellites and interplanetary spacecraft.
- Compare and contrast the general properties of the nine large planets.
- Describe the minor planets (asteroids).

CHAPTER 9

- Compare and contrast the general properties and surface conditions of Mercury, Venus, Earth, and Mars.
- Explain what is meant by "morning star" and "evening star."
- Compare and contrast the atmosphere of Mercury, Venus, Earth, Mars, and Jupiter.
- Describe conditions on Mars at the Viking 1 and 2 landing sites.
- Give two observations that indicate that water might once have been flowing on Mars.
- Compare and contrast the internal structure of Earth and Jupiter.
- Explain the theory of plate tectonics (continental drift) in relation to Earth's geological activity.
- List and give the current explanation for a famous feature visible in a small telescope for Venus, Mars, Jupiter, and Saturn.
- State some properties that Jupiter and Saturn have in common, and some common properties of Uranus and Neptune.
- Tell what is known about the moons of Mars, Jupiter, Saturn, Uranus, Neptune, and Pluto.
- Describe known properties of Pluto.

CHAPTER 10

- Explain the moon's appearance and motions in the sky.
- Compare the moon and Earth in diameter, mass, average density, and surface gravity.
- Describe the general surface features of the moon.
- Compare and contrast the moon and Earth in regard to geological activity and erosion of surface features.
- Explain the probable origin of lunar craters and maria.
- Describe surface conditions on the moon at the Apollo landing sites.
- Give the current picture of the moon's internal structure.
- Give the current theory of the moon's history.
- List some questions about the moon that remain to be answered.
- Describe the relative positions of Earth, moon, and sun during a solar and a lunar eclipse.

CHAPTER 11

- Describe the current theory of the origin and composition of comets.
- Explain in terms of the current theory of comet structure the changes in a comet's appearance as its distance from the sun changes.
- Explain the relationship between comets and meteor showers.

- Distinguish between a meteoroid, meteor, and meteorite.
- Give the composition and probable origin of meteorites.
- Explain why comets and meteorites are of interest to scientists.

CHAPTER 12

- Describe the molecular basis of Earth-life.
- Describe the evidence that indicates life may have evolved spontaneously from nonliving molecules on Earth.
- Describe a scientific theory of the origin and evolution of intelligent life on Earth.
- Describe the search for life on Mars.
- State the evidence for the existence of solar systems other than our own.
- List the factors involved in estimates of the statistical chances for extraterrestrial intelligent life.
- State the dominant current scientific view of interstellar voyages and UFOs.
- Describe several projects in which scientists have searched or are planning to search for extraterrestrial intelligence.

Useful References

MAGAZINES

Astronomy, 411 East Mason Street, Milwaukee, WI 53202
The Griffith Observer, Griffith Observatory, 2800 East Observatory Road, Los Angeles, CA 90027
Mercury, Astronomical Society of the Pacific, 1244 Noriega Street, San Francisco, CA 94122
National Geographic, 17th and M Streets, N.W., Washington, DC 20036
Natural History, Membership Services, Box 6000, Des Moines, IA 50340
Popular Astronomy, 270 Madison Avenue, New York, NY 10016
Science News, 1719 N Street, N.W., Washington, DC 20036
Scientific American, 415 Madison Avenue, New York, NY 10017
Sky and Telescope, 49-50-51 Bay State Road, Cambridge, MA 02138

OBSERVING GUIDES AND STAR ATLASES

A Field Guide to the Stars and Planets by Donald H. Menzel (Boston: Houghton Mifflin Co., 1975)
All About Telescopes by Sam Brown (Barrington, NJ: Edmund Scientific Co.)
The American Ephemeris and Nautical Almanac (Washington: U.S. Government Printing Office, yearly). Current information about sun, moon, planets, eclipses, and occultations.
Norton's Star Atlas and Reference Handbook, 16th ed. by Arthur P. Norton's successors (Cambridge, MA: Sky Publishing Corp., 1973)
The Observer's Handbook (Toronto, Canada: Royal Astronomical Society of Canada, yearly). Guide to current celestial events; many tables.
Sky Calendar (East Lansing, MI: Abrams Planetarium, Michigan State University, yearly)

GENERAL REFERENCE BOOKS

The author has used as principal reference works for this book:

Astrophysical Quantities, 3rd ed., by C. W. Allen (London: The Athlone Press of the University of London, 1973). This book has tables and lists that present the essential information of astrophysics.

Glossary of Astronomy and Astrophysics, 2nd ed., by Jeanne Hopkins (Chicago: University of Chicago Press, 1980). A technical glossary of the astronomical and astrophysical terms most commonly used in the *Astrophysical Journal.*

CAREER INFORMATION

A Career In Astronomy, The American Astronomical Society, Education Officer, Sharp Laboratory, University of Delaware, Newark, DE 19711.

Degree Programs in Physics and Astronomy in U.S. Colleges and Universities, American Institute of Physics, 335 East 45th Street, New York, NY 10017

Women In Science by Dinah L. Moché, American Association of Physics Teachers, Graduate Physics Building, State University of New York, Stony Brook, NY 11794. A multimedia package presenting information about six women scientists and their work. Includes written materials, audio cassette tapes, and 35 mm slides.

Selected Resources for Astronomy Materials

EDUCATIONAL MATERIALS, AUDIO-VISUAL AIDS

Books, color slides, astroposters, prints, and other astronomical audio-visual aids:

Astronomical Society of the Pacific, 1244 Noriega Street, San Francisco, CA 94122
Edmund Scientific Co., 101 East Gloucester Pike, Barrington, NJ 08007
Hansen Planetarium, 15 South State Street, Salt Lake City, UT 84111
Kitt Peak National Observatory, Public Information Office, P.O. Box 26732, Tucson, AZ 85726
MMI Corporation, 2950 Wyman Parkway, Baltimore, MD 21211
Sky Publishing Corporation, 49-50-51 Bay State Road, Cambridge, MA 02138

Other suppliers advertise in the monthly astronomy magazines listed on page 262.

Your nearest planetarium or science museum is a source of educational materials. For locations of planetariums and professional observatories:

Astronomical Directory by James Gall, Gall Publications:Toronto, 1978.
Planetarium Directory, Number 8, 1977, compiled by Bill Lazarus and Tom Fleming; for sale from Space Shop, Strasenburgh Planetarium, Box 1480, Rochester, NY 14603
U.S. Observatories: A Directory and Travel Guide, by H. T. Kirby-Smith; for sale from Sky Publishing Corporation, Cambridge, MA 02138

TEACHING MATERIALS

NASA films and space photography catalogues; available from NASA, Code FP, Audio Visual Branch, Public Information Division, 400 Maryland Avenue, S.W., Washington, DC 20546
NASA educational publications; for sale by Superintendent of Documents, U.S. Government Printing Office, Washington, DC 20402
National Air and Space Museum, Smithsonian Institution, Washington, DC 20560

Resource Letter EMAA-1: Educational Materials in Astronomy and Astrophysics by Richard Berendzen and David DeVorkin, for sale from American Association of Physics Teachers, Graduate Physics Building, State University of New York, Stony Brook, NY 11794

Resource Letter EMAA-2: Laboratory Experiences for Elementary Astronomy by Haym Kruglak, for sale from American Association of Physics Teachers, Graduate Physics Building, State University of New York, Stony Brook, NY 11794

Sky Publishing Corporation, 49-50-51 Bay State Road, Cambridge, MA 02138

OTHER BOOKS BY DINAH L. MOCHÉ

High School and Beyond

Life in Space, Ridge Press/A & W Visual Library: New York, 1979.
Radiation, Franklin Watts: New York, 1979.
Search for Life Beyond Earth, Franklin Watts: New York, 1978.

Young Readers:

The Astronauts, Random House: New York, Latest Printing 1981.
Mars, Franklin Watts: New York, 1978.
Magic Science Tricks, Scholastic Book Services: New York, 1977.
More Magic Science Tricks, Scholastic Book Services: New York, 1980.
My First Book about Space, Western Publishing: Racine, WI, 1982.
The Star Wars Question and Answer Book About Space, Random House: New York, Latest Printing 1981.
We're Taking an Airplane Trip, Western Publishing: Racine, WI, 1982.
What's Up There? Questions and Answers About Stars and Space, Scholastic Book Services: New York, Latest Printing 1981.

APPENDIX 1
Selected Constellations

Constellations

Name	Genitive	Abbreviation	Meaning	Reference RA	and Decl.
Andromeda	-dae	And	Chained Lady	1^h	+40°
Aquarius	-rii	Aqr	Water Bearer	23^h	−10°
Aquila	-lae	Aql	Eagle	19^h30^m	+ 5°
Aries	Arietis	Ari	Ram	2^h	+20°
Auriga	-gae	Aur	Charioteer	5^h30^m	+40°
Bootes	-tis	Boo	Herdsman	14^h30^m	+30°
Cancer	Cancri	Cnc	Crab	8^h30^m	+20°
Canes Venatici	Canum Venaticorum	CVn	Hunting Dogs	12^h30^m	+40°
Canis Major	-ris	CMa	Great Dog	7^h	−20°
Capricornus	-ni	Cap	Goat	21^h	−20°
Cassiopeia	-peiae	Cas	The Queen; also, Lady in the Chair	1^h	+60°
Cepheus	-phei	Cep	King (of Ethiopia)	22^h	+65°
Cetus	-ti	Cet	Whale	2^h	−10°
Coma Bernices	Comae-	Com	Bernice's Hair	13^h	+25°
Corona Borealis	-nae	CrB	Northern Crown	15^h30^m	+30°
Corvus	-vi	Crv	Crow	12^h30^m	−20°
Cygnus	-ni	Cyg	Swan; also, the Northern Cross	20^h	+40°
Draco	-conis	Dra	Dragon	18^h	+70°
Eridanus	-ni	Eri	The River Eridanus	3^h30^m	−20°
Gemini	-orum	Gem	Twins	7^h	+25°
Hercules	-lis	Her	Hercules	17^h	+35°
Hydra	-drae	Hyd	Sea-Serpent	11^h	−20°
Leo	Leonis	Leo	Lion	10^h30^m	+20°
Lepus	Leporis	Lep	Hare	5^h30^m	−20°
Libra	-ae	Lib	Scales	15^h	−15°
Lyra	-ae	Lyr	Harp	18^h30^m	+35°
Monoceros	-rotis	Mon	Unicorn	7^h	− 5°

Source: Material in Appendix 1 is reprinted by permission from *Fundamental Astronomy: Solar System and Beyond,* by Franklin W. Cole, © 1974 by John Wiley & Sons, Inc.

Constellations (*Continued*)

Name	Genitive	Abbreviation	Meaning	Reference RA and Decl.	
Ophiuchus	-chi	Oph	Serpent-Bearer	17^h	$0°$
Orion	-nis	Ori	Hunter	5^h30^m	$0°$
Pegasus	-si	Peg	Winged Horse	23^h30^m	$+20°$
Perseus	-sei	Per	Perseus	3^h30^m	$+45°$
Pisces	-cium	Pic	Fishes	23^h30^m	$+\ 5°$
Piscis Austrinis	-ini	PsA	Southern Fish	23^h	$-30°$
Sagittarius	-rii	Sgr	Archer	18^h30^m	$-30°$
Scorpio	-pionis	Sco	Scorpion	17^h	$-30°$
Serpens	-pentis	Ser	Serpent	16^h	$0°$
Taurus	-ri	Tau	Bull	4^h30^m	$+15°$
Triangulum	-li	Tri	Triangle	2^h	$+30°$
Ursa Major	-sae, -ris	UMa	Great Bear, also Big Dipper	11^h	$+60°$
Ursa Minor	-sae, -ris	UMi	Little Bear, also Little Dipper	15^h	$+70°$
Virgo	-ginis	Vir	Virgin	13^h	$-10°$
Vulpecula	-lae	Vul	Fox	19^h30^m	$+25°$

APPENDIX 2

The Physical and Astronomical Constants

*Velocity of light	c	= 299,792,458 meters per second
*Gravitation constant	G	= 6.672×10^{-11} m^3/kg · sec^2
Stefan-Bolzmann constant	σ	= 5.66956×10^{-5} erg cm^{-2} deg^{-4} sec^{-1}
Mass of electron	m_e	= 9.10956×10^{-28} gram
Mass of hydrogen atom	m_H	= 1.67352×10^{-24} gram
Mass of proton	m_p	= 1.672661×10^{-24} gram
*Astronomical unit	AU	= 149,597,870 kilometers
Parsec	pc	= 3.085678×10^{18} centimeters
		= 3.261633 light-years
Light-year	LY	= 9.460530×10^{17} centimeters
*Mass of sun	$M_\odot$	= 1.9891×10^{33} grams
*Radius of sun	**$R_\odot$	= 696,000 kilometers
Solar radiation	$L_\odot$	= 3.827×10^{33} ergs per second
*Mass of Earth	***$M_\oplus$	= 5.9742×10^{27} grams
*Equatorial radius of Earth	$R_\oplus$	= 6,378.140 kilometers
Direction of Galactic center	RA	= 264°.83, DEC = 28°.90 (1900)
*Ephemeris day	d_E	= 86,400 seconds
*Tropical year (equinox to equinox)		= 365.242199 ephemeris days
Sidereal year		= 365.256366 ephemeris days

*Adopted in 1976 by the International Astronomical Union at the 16th general assembly, at Grenoble, France.

**$\odot$ = of the sun

***$\oplus$ = of the earth

Source: Data reprinted by permission from *Astrophysical Quantities, 3rd edition,* © C. W. Allen, 1973 (Athlone Press, London).

APPENDIX 3

Measurements and Symbols*

METRIC CONVERSION FACTORS

Approximate Conversions to Metric Measures

Symbol	When You Know	Multiply by	To Find	Symbol
		Length		
in	inches	2.5	centimeters	cm
ft	feet	30.0	centimeters	cm
yd	yards	0.9	meters	m
mi	miles	1.6	kilometers	km
		Area		
in^2	square inches	6.5	square centimeters	cm^2
ft^2	square feet	0.09	square meters	m^2
yd^2	square yards	0.8	square meters	m^2
mi^2	square miles	2.6	square kilometers	km^2
		Mass (weight)		
oz	ounces	28.0	grams	g
lb	pounds	0.45	kilograms	kg
	short tons (2000 lb)	0.9	metric ton	t
		Volume		
gal	gallons	3.8	liters	L
ft^3	cubic feet	0.03	cubic meters	m^3
yd^3	cubic yards	0.76	cubic meters	m^3
		Temperature (exact)		
°F	degrees Fahrenheit	5/9 (after subtracting 32)	degrees Celsius	°C
°F	degrees Fahrenheit	5/9 (after adding 460)	degrees Kelvin	°K

*Courtesy U.S. Department of Commerce, National Bureau of Statistics.

ANGULAR MEASURE

A circle contains 360 degrees, or 360°.
1° contains 60 minutes of arc, or 60′.
1′ contains 60 seconds of arc, or 60″.

THE GREEK ALPHABET

Alpha	Α	α	Nu	Ν	ν
Beta	Β	β	Xi	Ξ	ξ
Gamma	Γ	γ	Omicron	Ο	o
Delta	Δ	δ	Pi	Π	π
Epsilon	Ε	ϵ	Rho	Ρ	ρ
Zeta	Ζ	ζ	Sigma	Σ	σ
Eta	Η	η	Tau	Τ	τ
Theta	Θ	θ, ϑ	Upsilon	Υ	υ
Iota	Ι	ι	Phi	Φ	ϕ, φ
Kappa	Κ	κ	Chi	Χ	χ
Lambda	Λ	λ	Psi	Ψ	ψ
Mu	Μ	μ	Omega	Ω	ω

SUN, MOON, PLANET SYMBOLS

☉	Sun	♂	Mars	♇	Pluto
☾	Moon	♃	Jupiter	●	new moon
☿	Mercury	♄	Saturn	☽	first quarter
♀	Venus	♅	Uranus	○	full moon
⊕	Earth	♆	Neptune	☾	last quarter

LIST OF THE CHEMICAL ELEMENTS

Element	Symbol	Atomic Number[1]
Hydrogen	H	1
Helium	He	2
Lithium	Li	3
Beryllium	Be	4
Boron	B	5
Carbon	C	6
Nitrogen	N	7
Oxygen	O	8
Fluorine	F	9
Neon	Ne	10
Sodium	Na	11
Magnesium	Mg	12
Aluminum	Al	13
Silicon	Si	14

[1]Atomic number is the number of protons in the nucleus of an atom of the element.

Element	Symbol	Atomic Number
Phosphorus	P	15
Sulfur	S	16
Chlorine	Cl	17
Argon	Ar	18
Potassium	K	19
Calcium	Ca	20
Scandium	Sc	21
Titanium	Ti	22
Vanadium	V	23
Chromium	Cr	24
Manganese	Mn	25
Iron	Fe	26
Cobalt	Co	27
Nickel	Ni	28
Copper	Cu	29
Zinc	Zn	30
Gallium	Ga	31
Germanium	Ge	32
Arsenic	As	33
Selenium	Se	34
Bromine	Br	35
Krypton	Kr	36
Rubidium	Rb	37
Strontium	Sr	38
Yttrium	Y	39
Zirconium	Zr	40
Niobium	Nb	41
Molybdenum	Mo	42
Technetium	Tc	43
Ruthenium	Ru	44
Rhodium	Rh	45
Palladium	Pd	46
Silver	Ag	47
Cadmium	Cd	48
Indium	In	49
Tin	Sn	50
Antimony	Sb	51
Tellurium	Te	52
Iodine	I	53
Xenon	Xe	54
Cesium	Cs	55

Element	Symbol	Atomic Number
Barium	Ba	56
Lanthanum	La	57
Cerium	Ce	58
Praseodymium	Pr	59
Neodymium	Nd	60
Promethium	Pm	61
Samarium	Sm	62
Europium	Eu	63
Gadolinium	Gd	64
Terbium	Tb	65
Dysprosium	Dy	66
Holmium	Ho	67
Erbium	Er	68
Thulium	Tm	69
Ytterbium	Yb	70
Lutecium	Lu	71
Hafnium	Hf	72
Tantalum	Ta	73
Tungsten	W	74
Rhenium	Re	75
Osmium	Os	76
Iridium	Ir	77
Platinum	Pt	78
Gold	Au	79
Mercury	Hg	80
Thallium	Tl	81
Lead	Pb	82
Bismuth	Bi	83
Polonium	Po	84
Astatine	At	85
Radon	Rn	86
Francium	Fr	87
Radium	Ra	88
Actinium	Ac	89
Thorium	Th	90
Protoactinium	Pa	91
Uranium	U	92
Neptunium	Np	93
Plutonium	Pu	94
Americium	Am	95
Curium	Cm	96

Element	Symbol	Atomic Number	Element	Symbol	Atomic Number
Berkelium	Bk	97	Mendeleevium	Mv	101
Californium	Cf	98	Nobelium	No	102
Einsteinium	E	99	Lawrencium	Lw	103
Fermium	Fm	100			

APPENDIX 4

The Messier Catalogue of Nebulae and Star Clusters

Messier no. (M)	NGC or (IC)††	Right ascension (1950)		Declination (1950)		Apparent visual magnitude	Description†
		h	m	°	′		
1	1952	5	31.5	+22	00	11.3	● Crab nebula in Taurus; remains of SN 1054
2	7089	21	30.9	− 1	02	6.4	Globular cluster in Aquarius
3	5272	13	39.8	+28	38	6.3	● Globular cluster in Canes Venatici
4	6121	16	20.6	−26	24	6.5	Globular cluster in Scorpio
5	5904	15	16.0	+ 2	17	6.1	Globular cluster in Serpens
6	6405	17	36.8	−32	10		Open cluster in Scorpio
7	6475	17	50.7	−34	48		Open cluster in Scorpio
8	6523	18	00.6	−24	23		● Lagoon nebula in Sagittarius
9	6333	17	16.3	−18	28	8.0	Globular cluster in Ophiuchus
10	6254	16	54.5	− 4	02	6.7	Globular cluster in Ophiuchus
11	6705	18	48.4	− 6	20		Open cluster in Scutum Sobieskii
12	6218	16	44.7	− 1	52	7.1	Globular cluster in Ophiuchus
13	6205	16	39.9	+36	33	5.9	● Globular cluster in Hercules
14	6402	17	35.0	− 3	13	8.5	Globular cluster in Ophiuchus
15	7078	21	27.5	+11	57	6.4	Globular cluster in Pegasus
16	6611	18	16.1	−13	48		● Open cluster with nebulosity in Serpens
17	6618	18	17.9	−16	12		Swan or Omega nebula in Sagittarius
18	6613	18	17.0	−17	09		Open cluster in Sagittarius
19	6273	16	59.5	−26	11	7.4	Globular cluster in Ophiuchus
20	6514	17	59.4	−23	02		Trifid nebula in Sagittarius

† A supplemental listing with descriptions, of the more interesting Messier objects follows the catalog entries.
†† NGC is the New General Catalogue of star clusters, galaxies and nebulae compiled by T. L. E. Dreyer in 1888. IC is the Index Catalogue, an extension of the NGC, compiled in the Harvard observatory annals, 1904, and revised in 1908.
● Especially fine observational targets.

Source: All material in Appendix 4 is reprinted by permission from *Fundamental Astronomy: Solar System and Beyond*, by Franklin W. Cole, © 1974 by John Wiley & Sons, Inc.

Messier no. (M)	NGC or (IC)††	Right ascension (1950)		Declination (1950)		Apparent visual magnitude	Description†
		h	m	°	′		
21	6531	18	01.6	−22	30		Open cluster in Sagittarius
22	6656	18	33.4	−23	57	5.6	Globular cluster in Sagittarius
23	6494	17	54.0	−19	00		Open cluster in Sagittarius
24	6603	18	15.5	−18	27		Open cluster in Sagittarius
25	(4725)	18	28.7	−19	17		Open cluster in Sagittarius
26	6694	18	42.5	− 9	27		Open cluster in Scutum Sobieskii
27	6853	19	57.5	+22	35	8.2	● Dumbbell planetary nebula in Vulpecula
28	6626	18	21.4	−24	53	7.6	Globular cluster in Sagittarius
29	6913	20	22.2	+38	21		Open cluster in Cygnus
30	7099	21	37.5	−23	24	7.7	Globular cluster in Capricornus
31	224	0	40.0	+41	00	3.5	● Andromeda galaxy
32	221	0	40.0	+40	36	8.2	● Elliptical galaxy; companion to M31
33	598	1	31.0	+30	24	5.8	Spiral galaxy in Triangulum
34	1039	2	38.8	+42	35		● Open cluster in Perseus
35	2168	6	05.7	+24	21		Open cluster in Gemini
36	1960	5	33.0	+34	04		Open cluster in Auriga
37	2099	5	49.1	+32	33		Open cluster in Auriga
38	1912	5	25.3	+35	47		Open cluster in Auriga
39	7092	21	30.4	+48	13		Open cluster in Cygnus
40		12	20	+59			● Close double star in Ursa Major
41	2287	6	44.9	−20	41		Loose open cluster in Canis Major
42	1976	5	32.9	− 5	25		● Orion nebula
43	1982	5	33.1	− 5	19		Northeast portion of Orion nebula
44	2632	8	37	+20	10		Praesepe; open cluster in Cancer
45		3	44.5	+23	57		● The Pleiades; open cluster in Taurus
46	2437	7	39.5	−14	42		Open cluster in Puppis
47	2478	7	52.4	−15	17		Loose group of stars in Puppis
48		8	11	− 1	40		Cluster of very small stars; not identifiable
49	4472	12	27.3	+ 8	16	8.5	Elliptical galaxy in Virgo
50	2323	7	00.6	− 8	16		Loose open cluster in Monoceros
51	5194	13	27.8	+47	27	8.4	Whirlpool spiral galaxy in Canes Venatici
52	7654	23	22.0	+61	20		Loose open cluster in Cassiopeia
53	5024	13	10.5	+18	26	7.8	Globular cluster in Coma Berenices
54	6715	18	51.9	−30	32	7.8	Globular cluster in Sagittarius
55	6809	19	36.8	−31	03	6.2	Globular cluster in Sagittarius

† A supplemental listing with descriptions, of the more interesting Messier objects follows the catalog entries.
†† NGC is the New General Catalogue of star clusters, galaxies and nebulae compiled by T. L. E. Dreyer in 1888. IC is the Index Catalogue, an extension of the NGC, compiled in the Harvard observatory annals, 1904, and revised in 1908.
● Especially fine observational targets.

Messier no. (M)	NGC or (IC)††	Right ascension (1950)		Declination (1950)		Apparent visual magnitude		Description†
		h	m	°	′			
56	6779	19	14.6	+30	05	8.7		Globular cluster in Lyra
57	6720	18	51.7	+32	58	9.0	●	Ring nebula; planetary nebula in Lyra
58	4579	12	35.2	+12	05	9.6		Spiral galaxy in Virgo
59	4621	12	39.5	+11	56	10.0		Spiral galaxy in Virgo
60	4649	12	41.1	+11	50	9.0		Elliptical galaxy in Virgo
61	4303	12	19.3	+ 4	45	9.6		Spiral galaxy in Virgo
62	6266	16	58.0	−30	02	7.3		Globular cluster in Scorpio
63	5055	13	13.5	+42	17	8.6	●	Spiral galaxy in Canes Venatici
64	4826	12	54.2	+21	57	8.5		Spiral galaxy in Coma Berenices
65	3623	11	16.3	+13	22	9.4		Spiral galaxy in Leo
66	3627	11	17.6	+13	16	9.0		Spiral galaxy in Leo; companion to M65
67	2682	8	48.4	+12	00			Open cluster in Cancer
68	4590	12	36.8	−26	29	8.2		Globular cluster in Hydra
69	6637	18	28.1	−32	24	8.0		Globular cluster in Sagittarius
70	6681	18	40.0	−32	20	8.1		Globular cluster in Sagittarius
71	6838	19	51.5	+18	39			Globular cluster in Sagitta
72	6981	20	50.7	−12	45	9.3		Globular cluster in Aquarius
73	6994	20	56.2	−12	50			Open cluster in Aquarius
74	628	1	34.0	+15	32	9.3		Spiral galaxy in Pisces
75	6864	20	03.1	−22	04	8.6		Globular cluster in Sagittarius
76	650	1	39.1	+51	19	11.4	●	Planetary nebula in Perseus
77	1068	2	40.1	− 0	12	8.9		Spiral galaxy in Cetus
78	2068	5	44.2	+ 0	02			Small emission nebula in Orion
79	1904	5	22.1	−24	34	7.5		Globular cluster in Lepus
80	6093	16	14.0	−22	52	7.5		Globular cluster in Scorpio
81	3031	9	51.7	+69	18	7.0	●	Spiral galaxy in Ursa Major
82	3034	9	51.9	+69	56	8.4		Irregular galaxy in Ursa Major
83	5236	13	34.2	−29	37	8.3		Spiral galaxy in Hydra
84	4374	12	22.6	+13	10	9.4		Elliptical galaxy in Virgo
85	4382	12	22.8	+18	28	9.3		Elliptical galaxy in Coma Berenices
86	4406	12	23.6	+13	13	9.2		Elliptical galaxy in Virgo
87	4486	12	28.2	+12	40	8.7		Elliptical galaxy in Virgo
88	4501	12	29.4	+14	42	9.5		Spiral galaxy in Coma Berenices
89	4552	12	33.1	+12	50	10.3		Elliptical galaxy in Virgo
90	4569	12	34.3	+13	26	9.6		Spiral galaxy in Virgo

† A supplemental listing with descriptions, of the more interesting Messier objects follows the catalog entries.
†† NGC is the New General Catalogue of star clusters, galaxies and nebulae compiled by T. L. E. Dreyer in 1888. IC is the Index Catalogue, an extension of the NGC, compiled in the Harvard observatory annals, 1904, and revised in 1908.
● Especially fine observational targets.

Messier no. (M)	NGC or (IC)††	Right ascension (1950)		Declination (1950)		Apparent visual magnitude	Description†
		h	m	°	′		
91		omitted					
92	6341	17	15.6	+43	12	6.4	Globular cluster in Hercules
93	2447	7	42.4	−23	45		Open cluster in Puppis
94	4736	12	48.6	+41	24	8.3	Spiral galaxy in Canes Venatici
95	3351	10	41.3	+11	58	9.8	Barred spiral galaxy in Leo
96	3368	10	44.1	+12	05	9.3	Spiral galaxy in Leo
97	3587	11	12.0	+55	17	11.1	Owl nebula; planetary nebula in Ursa Major
98	4192	12	11.2	+15	11	10.2	Spiral galaxy in Coma Berenices
99	4254	12	16.3	+14	42	9.9	Spiral galaxy in Coma Berenices
100	4321	12	20.4	+16	06	9.4	Spiral galaxy in Coma Berenices
101	5457	14	01.4	+54	36	7.9	Spiral galaxy in Ursa Major
102		omitted					
103	581	1	29.9	+60	26		Open cluster in Cassiopeia
104	4594	12	37.4	−11	21	8.3	Spiral galaxy in Virgo
105	3379	10	45.2	+13	01	9.7	Elliptical galaxy in Leo
106	4258	12	16.5	+47	35	8.4	Spiral galaxy in Canes Venatici
107	6171	16	29.7	−12	57	9.2	Globular cluster in Ophiuchus

† A supplemental listing with descriptions, of the more interesting Messier objects follows the catalog entries.
†† NGC is the New General Catalogue of star clusters, galaxies and nebulae compiled by T. L. E. Dreyer in 1888. IC is the Index Catalogue, an extension of the NGC, compiled in the Harvard observatory annals, 1904, and revised in 1908.
● Especially fine observational targets.

Selected Messier Objects

M Number	Constellation	Right Ascension		Declination		Description
		h	m	°	′	
M103	Cassiopeia	1	29.8	+60	26	A fine open cluster, 1° field; contains a red star.
M31	Andromeda	0	40.0	+41.0		The Great nebula in Andromeda, a spiral galaxy visible as a hazy spot to the unaided eye (Figure 5.12).
M33	Triangulum	1	31	+30	24	Very large, faint, ill-defined spiral galaxy. Use low power on a clear night. Spiral arms visible only in photographs.
M42	Orion	5	32.9	− 5	25	The great Orion nebula visible as a hazy star in Orion's sword Best seen with low power. Probably the most beautiful of the gaseous nebulae accessible to small telescopes (Figure 4.1).
M1	Taurus	5	31.5	+21	59	The Crab nebula (Figure 4.10). Faintly visible as an oval patch; the serrated edge is visible only in large telescopes.
M44	Cancer	8	37.2	+20	10	Praesepe, The Beehive; a large, scattered open cluster almost visible with the unaided eye. Best seen with low power.
M46	Puppis	7	39.5	−14	42	A beautiful cluster about the size of the full moon. The irregular planetary nebula NGC 2438 is on northern edge.
M3	Canes Venatici	13	39.9	+28	38	Beautiful, bright, condensed globular cluster. Best seen with a 6-in. or larger telescope at high power.
M13	Hercules	16	39.9	+36	33	Probably the grandest globular cluster of all (Figure 5.5). Visible with binoculars.
M57	Lyra	18	52.0	+32	58	The Ring nebula; oval in shape resembling a smoky doughnut. Magnifies well (Figure 4.8).
M27	Vulpecula	19	57.4	+22	35	The Dumbbell nebula; elliptical in appearance, with fairly luminous notches.

APPENDIX 5

The Satellites of the Solar System

PLANET	SATELLITE	DISTANCE FROM PLANET[a] (km)	DIAMETER (km)	REVOLUTION PERIOD[b] *d*ays, *h*ours, *m*inutes	DISCOVERY
EARTH	Moon	384,400	3,476	27d 7h 41m	
MARS	Phobos	9,379	21	7h 39m	Hall, 1877
	Deimos	23,459	15	1d 6h 18m	Hall, 1877
JUPITER	XVI 1979J3	(127,800)	(40)	7h 4m	Synnot, 1980
	XIV 1979J1	128,000	(30-40)	7h 8m	Jewitt, 1979
	V Amalthea	181,500	265	11h 57m	Barnard, 1892
	XV 1979J2	223,000	(80)	16h 16m	Synnott, 1980
	I Io	422,000	3,632	1d 18h 28m	Galileo, 1610
	II Europa	671,400	3,126	3d 13h 14m	Galileo, 1610
	III Ganymede	1,071,000	5,276	7d 3h 43m	Galileo, 1610
	IV Callisto	1,884,000	4,820	16d 16h 32m	Galileo, 1610
	XIII Leda	11,094,000	(7)	238.7d	Kowal, 1974
	VI Himalia	11,487,000	170	250.57d	Perrine, 1904
	X Lysithea	11,861,000	(14)	263.55d	Nicholson, 1938
	VII Elara	11,747,000	80	259.65d	Perrine, 1905
	XII Ananke	21,250,000	(14)	631d	Nicholson, 1951
	XI Carme	22,540,000	(14)	692d	Nicholson, 1938
	VIII Pasiphae	23,510,000	(16)	739d	Melotte, 1908
	IX Sinope	23,670,000	(14)	758d	Nicholson, 1914
SATURN	1980S28	137,300	30	14.45h	Voyager 1, 1980
	1980S27	139,400	220	14.71h	Voyager 1, 1980
	1980S26	141,700	200	15.09h	Voyager 1, 1980
	1980S3	151,400	90 × 40	16.66h	Voyager 1, 1980
	1981S1	157,500	100 × 90	16.67h	Voyager 1, 1980
	1981S12	185,000	(10)	22.61h	Synott, 1982*
	Mimas	188,200	390	0.96d	Herschel, 1789
	Enceladus	240,200	500	1.39d	Herschel, 1789
	1981S6	295,000	(20)	1.89d	Synott, 1982*
	1980S13	296,600	(30 × 40)	1.91d	Un. of Az., 1981
	Tethys	296,600	1,050	1.91d	Cassini, 1684
	1980S25	296,600	(30 × 40)	1.91d	Westphall, 1981
	1981S10	350,000	(15)	2.44d	Synott, 1982
	1981S7	375,000	(15)	2.73d	Synott, 1982*
	1980S6	378,600	(160)	2.74d	Voyager 1, 1980
	Dione	379,100	1,120	2.76d	Cassini, 1684
	1981S9	470,000	(20)	3.8d	Synott, 1982
	Rhea	527,800	1,530	4.53d	Cassini, 1672
	Titan	1,221,400	5,140	15.94d	Huygens, 1655
	Hyperion	1,502,300	290	21.74d	Bond, 1848
	Iapetus	3,559,400	1,440	79.24d	Cassini, 1671
	Phoebe	10,583,200	(160)	406.49d	Pickering, 1898
URANUS	Miranda	129,800	(300)	1.41d	Kuiper, 1948
	Ariel	190,900	—	2.52d	Lassell, 1851
	Umbriel	266,000	(400)	4.14d	Lassell, 1851
	Titania	436,000	(1,700)	8.71d	Herschel, 1787
	Oberon	583,400	—	13.46d	Herschel, 1787
NEPTUNE	Triton	355,550	4,400	5.88d	Lassell, 1846
	Nereid	5,567,000	(500)	359.88d	Kuiper, 1949
PLUTO	Charon	~17,000	—	~6.39d	Christy, 1978

a Semi-major axis
b Sidereal period
Parentheses indicate value is estimated
* Data obtained by Voyager in 1981; discovered in 1982

Source: Adapted and selected from *National Aeronautics and Space Administration* public information.

Index

CAMBRIA COUNTY LIBRARY
JOHNSTOWN, PA. 15901

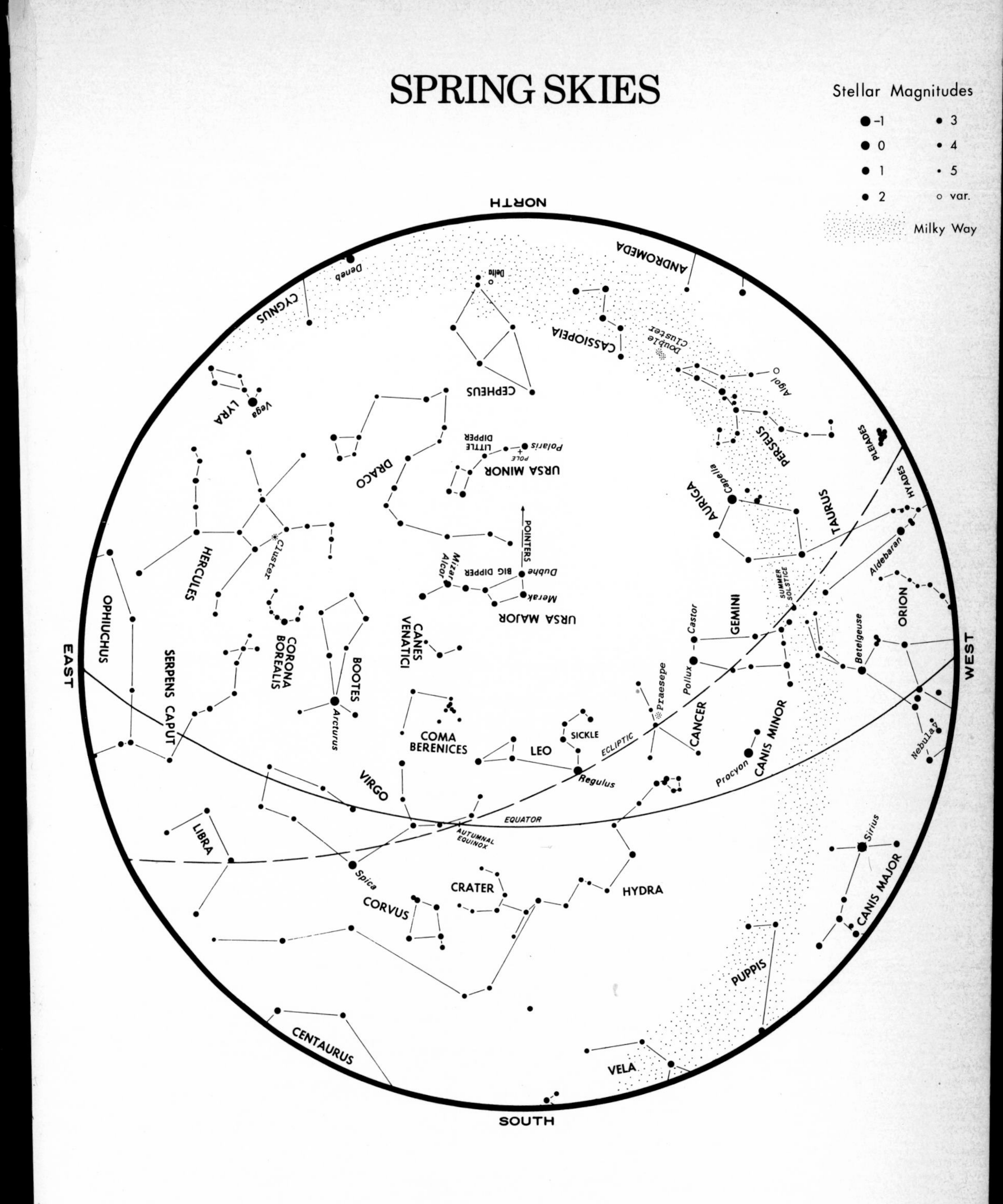

TIME SCHEDULE

Chart by George Lovi

Late March	11 p.m.	Early May	8 p.m.
Early April	10 p.m.	Late May	7 p.m.
Late April	9 p.m.	Early June	6 p.m.

STANDARD TIME

SUMMER SKIES

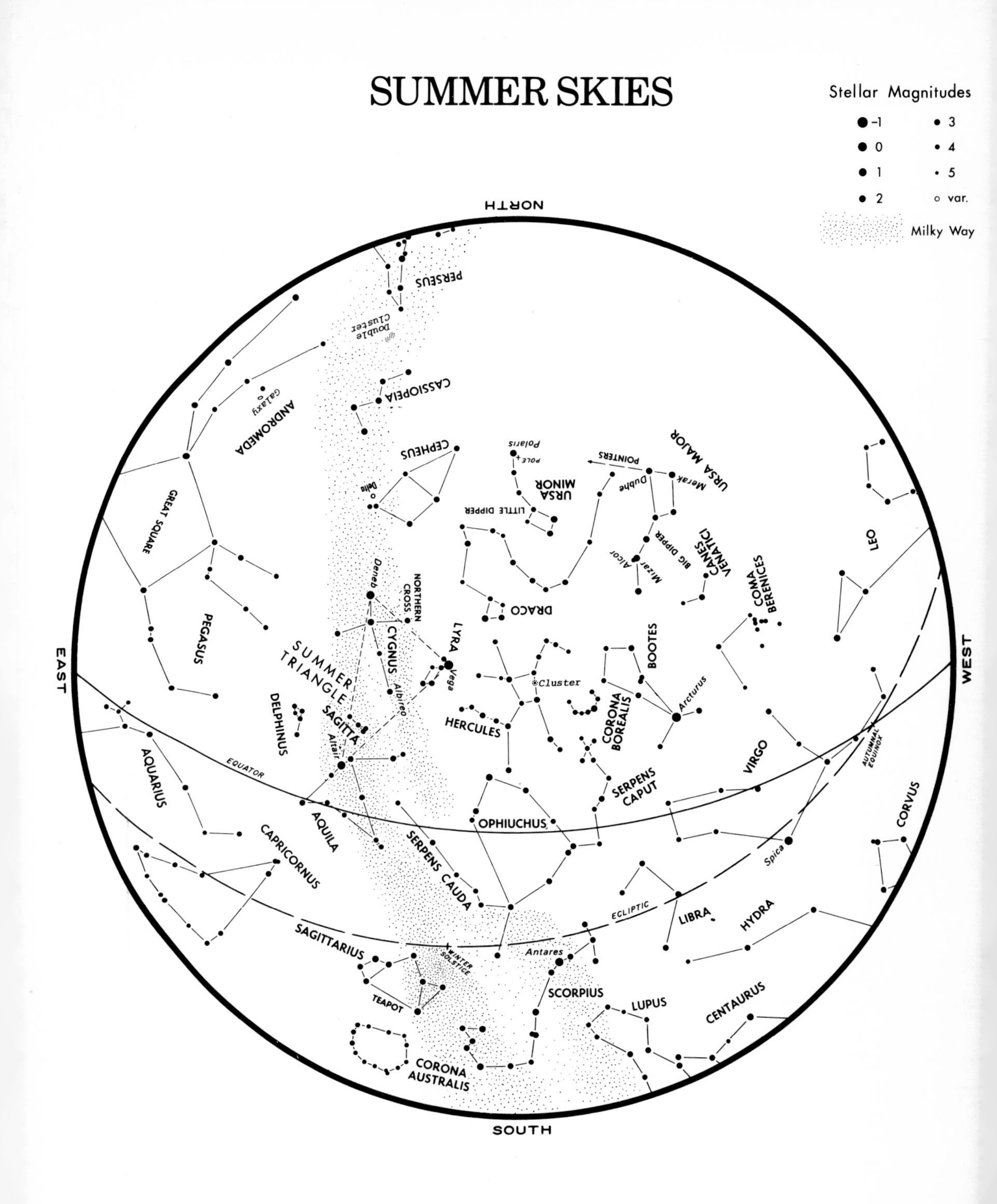

TIME SCHEDULE

Chart by George Lovi

Late June	11 p.m.	Early August	8 p.m.
Early July	10 p.m.	Late August	7 p.m.
Late July	9 p.m.	Early September	6 p.m.

STANDARD TIME